STRENGTHENING DETERRENCE
NATO and the Credibility of
Western Defense in the 1980s

THE ATLANTIC COUNCIL'S WORKING GROUP
ON THE CREDIBILITY OF THE NATO DETERRENT

KENNETH RUSH and
BRENT SCOWCROFT, *Cochairmen*

JOSEPH J. WOLF, *Rapporteur and Editor*

BALLINGER PUBLISHING COMPANY
Cambridge, Massachusetts
A Subsidiary of Harper & Row, Publishers, Inc.

International Standard Book Number: 0-88410-868-6

Library of Congress Catalog Card Number: 81-20546

Printed in the United States of America

Library of Congress Cataloging in Publication Data

Main entry under title:

The Credibility of the NATO deterrent.

  Includes index.
  1. North Atlantic Treaty Organization—Addresses.
essays, lectures. I. Wolf, Joseph J. II. Atlantic
Council of the United States. III. Title: Credibility
of the N.A.T.O. deterrent.
UA646.3.C73   355'.031'091821   81-20546
ISBN 0-88410-868-6   AACR2

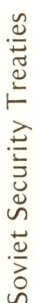

Soviet Security Treaties

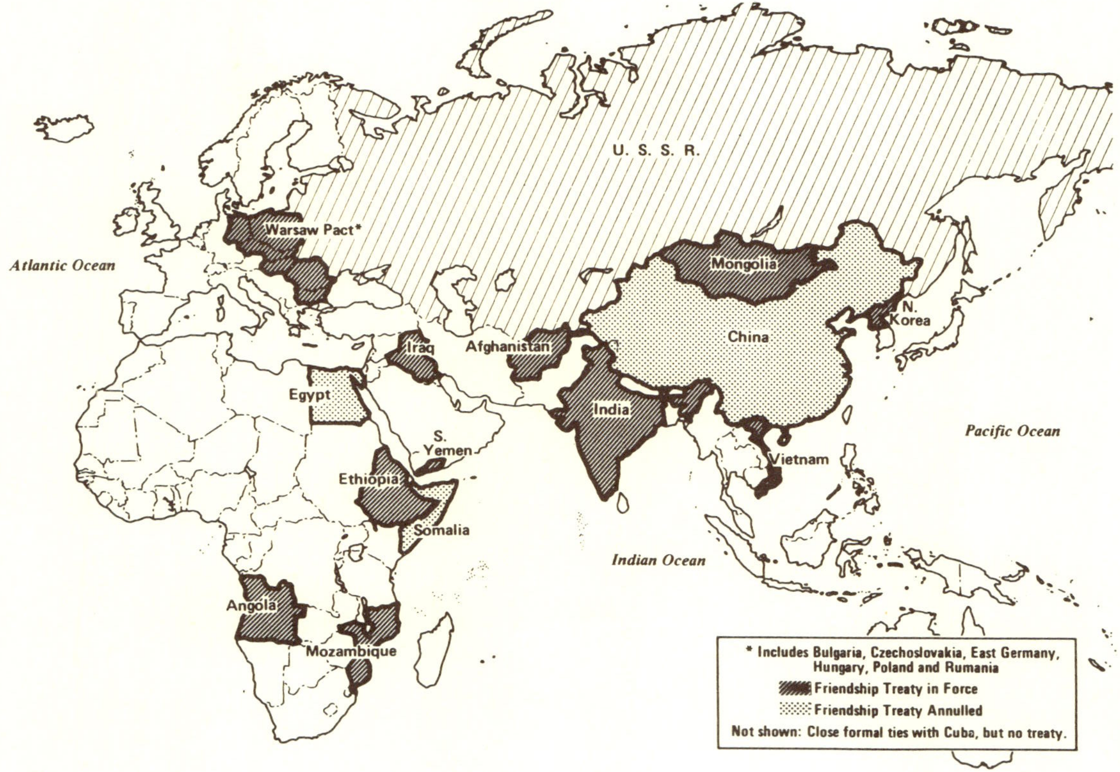

Note: Permission granted by the Library of Congress to reproduce this map from the publication entitled "U.S.–Soviet Military Balance Concepts and Capabilities 1960–1980."

LIBRARIES
UNIVERSITY OF MAINE
AT ORONO

RAYMOND H. FOGLER LIBRARY

ORONO

STRENGTHENING DETERRENCE

To the Memory of James S. McDonnell, who served as a member of the Board of Directors of the Atlantic Council for many years, with deep appreciation for his able and generous support of the Council and for his many valuable contributions to the development of a stronger NATO, a better United Nations, and a more effective world order.

CONTENTS

List of Figures and Tables   ix

List of Members of the Atlantic Council Working Group on the Credibility of the NATO Deterrent   xi

Foreword — *Kenneth Rush and Brent Scowcroft*   xv

Introduction
— *Joseph J. Wolf*   1

Chapter 1
Policy Paper. The Credibility of the NATO Deterrent: Bringing the NATO Deterrent Up to Date
— *The Atlantic Council's Working Group on The Credibility of the NATO Deterrent*   7

Chapter 2
The USSR and the Western Alliance
— *William G. Hyland*   51

CONTENTS

Chapter 3
U.S.-Allied Relations: The Current Crisis in Historical Perspective
— *Robert E. Osgood* — 73

Chapter 4
The Military Balance
— *A Staff Study* — 107

Chapter 5
Problems of Readiness, Reinforcement, and Resupply
— *George C. Blanchard, Isaac C. Kidd, Jr., and John W. Vogt* — 135

Chapter 6
The Impact on NATO of Security Requirements outside the Treaty Area
— *Jeffrey Record* — 163

Chapter 7
Concepts and Capabilities
— *Russell E. Dogherty* — 191

Chapter 8
Resources and Requirements
— *Roy A. Werner* — 211

Chapter 9
Priorities for the Future
— *George M. Seignious II* — 249

Glossary — 261

Index — 263

LIST OF FIGURES AND TABLES

Figures

4-1	Soviet Forces in Central Europe	113
4-2	NATO Forces in Central Europe	115
5-1	NATO Command Structure	143

Tables

5-1	Shortages of Noncommissioned and Petty Officers (E5-E9), End of Fiscal Year 1981	138
5-2	Shortages of U.S. Reserve Forces, End of Fiscal Year 1981	140
6-1	Tentative Rapid Deployment Joint Task Force Composition, 1981	182
8-1	Defense Expenditures and National Economies, 1979	216
8-2	U.S. Net Import of Selected Minerals and Metals, Percentage Consumption, 1979	226
8-3	Investment and Productivity in the United States and Other Industrial Nations, 1960-1976	235
8-4	Declining Percentage of U.S. Markets Held by American Producers	235
8-5	Shortfalls in U.S. Military Personnel in the Event of Mobilization	237

MEMBERS OF THE ATLANTIC COUNCIL WORKING GROUP ON THE CREDIBILITY OF THE NATO DETERRENT

COCHAIRMEN

Kenneth Rush, chairman, Atlantic Council; former deputy and acting secretary of state, deputy secretary of defense, and ambassador to France and Germany.

Brent Scowcroft, former assistant to the president of the United States for national security affairs.

PROJECT DIRECTOR

Francis O. Wilcox, director general, Atlantic Council; former assistant secretary of state and dean of the School of Advanced International Studies, The Johns Hopkins University.

RAPPORTEUR

Joseph J. Wolf, former minister, U.S. delegation to NATO.

MEMBERS

Theodore C. Achilles, vice chairman, Atlantic Council; former counselor of the Department of State and ambassador to Peru.

George Blanchard, former commander-in-chief, U.S. Army, Europe, and commander, Central Army Group.

Robert Bowie, professor of government, Harvard University; former assistant director of the U.S. Central Intelligence Agency and director, policy planning staff, U.S. Department of State.

MEMBERS OF THE ATLANTIC COUNCIL WORKING GROUP

Richard Burt, Journalist, *New York Times*; participated prior to appointment as director, office of politico-military affairs, U.S. Department of State.

Thomas A. Callaghan, Jr., president, Ex-Im Tech, and associate, Georgetown Center for Strategic and International Studies.

Arthur Cyr, vice president and program director, Chicago Council on Foreign Relations.

Russell E. Dougherty, formerly commanding U.S. general, Strategic Air Command, and chief of U.S. staff, SHAPE.

Robert F. Ellsworth, president, Robert Ellsworth & Co.; former deputy secretary of defense and U.S. ambassador to NATO.

Andrew J. Goodpaster, former superintendent, U.S. Military Academy and supreme allied commander, Europe.

Lincoln Gordon, senior fellow, Resources for the Future; former assistant secretary of state and ambassador to Brazil; participated prior to appointment as member of the senior review panel, U.S. Central Intelligence Agency.

Joseph W. Harned, deputy director general, Atlantic Council.

Livingston Hartley, author (deceased).

John D. Hickerson, former assistant secretary of state and ambassador to Finland and the Philippines.

Joseph Harsch, journalist, *The Christian Science Monitor.*

Martin J. Hillenbrand, director general, Atlantic Institute for International Affairs; former U.S. ambassador to Germany.

Ann Holick, former associate professor and executive director of the Ocean Policy Project, School of Advanced International Studies, The John Hopkins University; participated prior to appointment as director, policy assessment staff, U.S. Department of State.

William G. Hyland, Carnegie Endowment for International Peace; former deputy assistant to the president of the United States for national security affairs.

U. Alexis Johnson, former chief, U.S. delegation to Strategic Arms Limitation Treaty negotiations and under secretary of state.

George Jones, journalist, *U.S. News and World Report.*

Julius Katz, former assistant secretary of state for economic and business affairs.

Isaac Kidd, Jr., former supreme allied commander, Atlantic.

Lane Kirkland, president, AFL-CIO.

Clinton E. Knox, former ambassador to Haiti and the Republic of Dahomey (deceased).

Lyman L. Lemnitzer, former supreme allied commander, Europe.

Jay Lovestone, AFL-CIO and ILGWU Consultant on National Affairs.

Robert McFarlane, former special assistant to the president of the United States, for national security affairs; participated prior to appointment as counselor, U.S. Department of State.

George McGhee, former under secretary of state and ambassador to Germany and Turkey.

Henry Nau, associate professor of political science, George Washington University; participated prior to appointment to National Security Council staff.
Paul H. Nitze, former secretary of the navy and deputy secretary of defense.
Robert E. Osgood, former dean of School of Advanced International Studies, The Johns Hopkins University.
William L. Read, former commander, Naval Surface Force, U.S. Atlantic fleet.
Jeffrey Record, senior fellow, Institute for Foreign Policy Analysis.
Stanley Resor, attorney, Debevoise, Plimpton, Lyons and Gates; former secretary of the Army and U.S. Ambassador to the MBFR negotiations.
Eugene V. Rostow, professor of law, Yale University; former under secretary of state; participated prior to appointment as director, Arms Control and Disarmament Agency.
Thomas C. Schelling, professor of political economy, Harvard University.
George M. Seignious, II, former president of the Citadel, deputy assistant secretary of defense, and director, Arms Control and Disarmament Agency.
Helmut Sonnenfeldt, guest scholar, Brookings Institution; former counselor, U.S. Department of State.
Timothy W. Stanley, president, International Economic Policy Association.
Robert Strausz-Hupe, author; former ambassador to NATO, Sweden, Belgium, and Ceylon.
Leonard Sullivan, Jr., defense policy consultant; former assistant secretary of defense.
John Vogt, former commander-in-chief, Pacific Air Forces, and commander, Allied Air Forces, Central Europe.
Roy A. Werner, special assistant to the president, Aerojet Ordnance; former Principal Deputy Assistant Secretary of the Army.
Thomas Wolfe, senior staff member, The Rand Corporation.

PROJECT ASSISTANTS

Martha C. Finley
Kathryn M. Coulter

FOREWORD

An increasingly dangerous military imbalance between East and West in Europe, the loss of Western nuclear superiority, Communist exploitation of mounting instability in countries outside the NATO area, and the lack of a cohesive response on the part of the Western allies all combine to raise serious and profound doubts about the prospects of the members of the North Atlantic Treaty Organization.

That is why the Atlantic Council in 1979 decided to undertake the study of the credibility of the NATO deterrent reported in this book. We were fortunate indeed to attract to a special working group for this purpose (see list of names) some of the most knowledgeable experts available, both military and civilian. Death has since deprived us of the long experience and wise counsel of Livingston Hartley and Clinton E. Knox, who had devoted their lives to the cause of peace and freedom and who are deeply missed.

After general exploration of the problem at its opening meetings, the working group based its discussions on a series of background papers, which were subsequently developed and expanded by their authors for this book. These studies and the resulting policy paper are, we believe, of outstanding quality and value and will commend themselves to all who turn these pages.

This is the fourth recent study by the Atlantic Council of the security concerns of the Atlantic allies. In 1977 our policy paper on

The Growing Dimensions of Security drew attention to the global nature of the threat and the nonmilitary as well as military challenges to security. In 1978 *Securing the Seas* reflected a two years' study of the Soviet naval challenge and the options open to the Western alliance. In 1980 the problems of security in the Third World were analyzed in *After Afghanistan: The Long Haul*.

The present inquiry also examines the significance for NATO defenses of threats outside the NATO area in light of the steadily growing Soviet military power. The reader is reminded of the modern uses of military power, by which major states seek to hide behind activities of surrogate states for intimidation, for the support of insurrection, and for the overthrow of established governments by means short of war. Growing inadequacies in NATO's defense are identified and the significance for NATO of the loss of Western nuclear superiority explored. Ways to compensate for new demands on the allies strength outside the NATO area are sought. Finally, the weaknesses within the alliance that endanger the continued solidarity and common purpose of this essential coalition are described.

Copies of the final draft of the policy paper, Chapter 1, were made available to Dr. Joseph M.A. Luns, Secretary General of NATO; General Bernard W. Rogers, Supreme Allied Commander, Europe; and Admiral Harry D. Train II, Supreme Allied Commander, Atlantic. The views expressed herein, however, are solely the responsibility of the working group and the respective authors and do not necessarily reflect those of any of the NATO authorities.

Special thanks go to the corporations and foundations whose generous help made this study possible. Partial support for the project was received from the U.S. Departments of State and Defense. The opinions, conclusions, and recommendations expressed do not necessarily reflect those of either department. The policy paper reflects the views of the working group and not necessarily those of the Atlantic Council as a whole. Although Chapter 1 sets forth the general views of the working group, it should not be inferred that every member supports every statement therein. The other chapters, background for the policy, reflect the views of the respective authors, though they of course reflect in varying degrees the views of the working group as a whole.

Support for the overall Atlantic Council program of which this effort is a part has been made available by a number of American foundations, corporations, labor unions, and individuals, We are most grateful for this support, as well as for the extraordinary contributions of time and talent by the working group members.

The most important message of these studies is that the allies must *jointly* solve the problems that face us all. There is no room in today's world for dealing with each other at arm's length. Only by the closest, frankest, and most open discussion of our problems can we hope to find the common solutions that history now demands of us.

Kenneth Rush   **Brent Scowcroft**
Cochairmen of the Atlantic Council's Working Group on the Credibility of the NATO Deterrent

INTRODUCTION
Joseph J. Wolf

Comparing the world in the last quarter of the twentieth century thus far with the world of this century's second and third quarters, it is clear that one aspect of maintaining peace has changed little. The question of security still remains how nations dedicated to the rights of the individual under independent responsible governments can protect their liberties against expansionist totalitarianism.

    The concept of security has been redefined in recent years and quite properly so, to encompass many problems previously perceived to be only social or economic. Widespread hunger, rapidly multiplying population, energy shortage, unemployment, violations of human rights and social justice, and tension in the relationship between industrialized and developing countries profoundly affect the security of individual nations, regions, and indeed, the entire world. No security policy that fails to treat these elements of modern international life can be considered complete. Though vital, however, these elements are outside the immediate scope of the present book, which focuses on the more limited but equally vital question of the adequacy of the NATO deterrent under contemporary conditions.

    A major change affecting security lies in the fact that the relations between nations have altered dramatically since the Second World War. Even though the issue for the non-Communist nations created by the hegemonic thrust of the USSR may not be new, the significance of these changes for the Western allies is more profound than is at times understood, as explained in the following paragraphs:

- The colonial empires of the past have been replaced by a multitude of diverse new and growing independent nations. The power of the former colonial masters has been deeply constrained by this development; the new nations, for the most part pursuing a policy of nonalignment and intensely individual nationalism, have by no means replaced the lost power. The apparent stability of the colonial system has been replaced by endemic instability, as the new nations have sought through trial and error, and all too often through violence, to find social and political systems suitable to their needs.

- Despite the loss of their colonial domains, the countries of Western Europe have emerged as one of the major centers of economic strength. The gross national product per capita of four of them exceeds that of the United States. Trade within the European community likewise has burgeoned. A vastly improved standard of living has flourished out of the rubble of World War II.

- North Asia, too, recovering from two devastating wars, provides the economic miracles of Japan, South Korea, and Taiwan, which are now being emulated in Hong Kong and Singapore; and the Chinese giant is awakening.

- Having endured enormous sacrifice in World War II, the Soviet Union has indeed emerged as a superpower. In addition to attaining general nuclear parity with the United States, by consistently heavy defense production at the expense of consumer goods over many years, the Soviet Union has made itself into a modern, first-rate military power with truly global reach and the capability to attack as well as defend in the European military theater. The advent of the Peoples Republic of China as a power on the world scene, on the other hand, can give the USSR little comfort as it plans for the future.

- Despite serious internal economic problems, the USSR has markedly improved its economic relations with Western Europe as part of its policy of stabilizing its Western front. The confirmation of the role of Eastern Europe as a buffer for the Soviet Union in the series of agreements of which the Helsinki accords were the latest added to the impact of détente in Europe on Western European perceptions.

- The United States, at one time economically unchallengeable, militarily predominant, and clearly superior in nuclear capability is now emerging from a period in which its once unquestioned leadership has undergone some hard knocks. A weakened economy and currency, a turning aside from global responsibility as a result of Vietnam War, the loss of nuclear superiority, and a prolonged period of inadequate investment in defense have combined to undermine the impression of an unassailable America.

- The capability of the Soviet Union, alone or with the help of radical elements of the region, to interfere with the flow of oil from the Middle East to the Western nations has created an additional strain on the resources of the West.

- Long-standing relationships among the industrialized nations have also been affected. For example, trade between Europe and the United States is no longer as significant as it was before the rise in East Asian trade with the United States and trade among the nations of Europe, both Eastern and Western.

All of these factors affect significantly what can be done to keep NATO the force for peace it has been for so long and increase the difficulties. Most important, they emphasize the interdependence of the Western nations and the need more than ever for collective, rather than individual security.

The United States no longer towers over its associates and allies to the extent it once did—or over its enemies. But the United States can more than hold its own in the world and together with its allies, who are stronger than ever before, can surely do what is needed for safety.

What does all this imply for the NATO alliance? Much remains apparently the same as in earlier years. The need for the alliance has not changed. Neither the policies nor the deeds of the Soviet Union give any cause for comfort that its expansionist thrust has any limits other than those of pragmatism. The use of raw Soviet power in Afghanistan has been a caution to be heeded. The continuing need for the alliance stems from the dangers that would still be inherent in any significant imbalance of power, putting the Soviet Union, vastly larger, stronger, and better armed than any one of the nations of Western Europe, in a position to influence and dominate their policies and future without armed attack. With Europe under Soviet influence or control, there would be a fundamental change in the power structure of the world, in turn isolating the United States and endangering its future economically, politically, and militarily. The repercussions would be worldwide.

The *task* of the alliance is fundamentally the same. Now called "deterrence and détente," it was more simply put in 1949, in words less controversial and less subject to misunderstanding, by Senator Arthur Vandenberg, one of the godfathers of collective security: "to stop another war before it starts, and to avoid appeasements which are but surrender on the installment plan."

The purpose of the alliance likewise remains basically unchanged. It is to protect a way of life founded on respect for both the liberty of the individual and the rights of others. As restated at the NATO heads of government meeting of May 11, 1977, it is: "to safeguard the independence and security of its members, enabling them to promote the values of democracy and respect for human rights and individual freedom, and to make possible a lasting state of peace."

No more has the structure of the alliance changed. It is still a free alliance of sovereign nations, with all of the strengths and weaknesses inherent in that sort of structure. It thus reflects neither a single homogeneous policy on the part of its members, nor even identical though separate viewpoints and policies on all issues. Rather, it has reflected the individuality and diversity as well as the common interests of the nations that make it up. It is held together, not by the hegemonic domination of the strongest member, but by the continuing conviction of every independent member, the weakest as well as the strongest, that its members' own national interests are well served thereby. Its ties perforce have had to be elastic and flexible. They are less likely to snap under stress as would those of some more rigid form of association, but they are consequently subject to concern that they might become too loose to be effective.

The nature of the alliance has not changed either. Notwithstanding attempts to categorize it as a purely military institution, it has always encompassed political as well as military goals. The concept of deterrence of war itself combines the two goals: by being militarily strong the alliance serves the political goal of not having to use that strength. Military power has been the most essential part of a broad insurance policy against the long shadow of fear cast by the potential for external domination, as well as against the more remote danger of armed attack itself.

Thus adequate military strength is not only essential for deterrence and defense; it is also essential for avoiding the more likely threat of piecemeal appeasement. It provides the confidence that permits firmness against external pressures short of the use of military force. Yet it is by no means the whole story. As Senator Vandenberg warned: "Unless this Treaty [NATO] becomes far more than a military alliance, it will be at the mercy of the first plausible Soviet peace offensive."

These principles of collective security remain constant over the years, but applying them to changed circumstances requires a willingness to adapt to current needs and to break free of institutional and historic tradition to the extent tradition fails to respond to modern conditions. The West must find the wit to turn the past into prologue, to recognize that changes in the international relationships between nations have occurred, and to adapt its ways of doing things to respond to those developments.

1 POLICY PAPER. THE CREDIBILITY OF THE NATO DETERRENT
Bringing the NATO Deterrent Up to Date

The Atlantic Council's Working Group on the Credibility of the NATO Deterrent

THE ATLANTIC ALLIANCE AND TODAY'S CHALLENGES

The fixed coastal guns of Singapore, pointed seaward, offered no defense to the overland attack from the opposite direction before which Singapore fell in World War II. The NATO nations must take heed of that lesson. They must make sure that their defenses cover all of today's manifold threats and challenges.

For the Atlantic Allies cannot escape these developments. Any illusions that such an escape was possible disappeared with the threat to access to Middle Eastern and Persian Gulf oil. That threat posed almost as serious a challenge to the security and way of life of the Atlantic allies as would armed attack across the Oder–Neisse line, yet paradoxically increased rather than reduced the differences among the allies.

So at the very time that the problems of security confronting the Atlantic allies in today's world are far more multi-dimensional and troublesome than ever before, the allies are faced with the prospects of more discouraging disarray than this troubled but resilient alliance has heretofore known.

The problems of security are now global as well as regional. They are now political and economic as well as military. They encompass relations with non-aligned as well as aligned nations. And because of their very difficulty and complexity, they have produced and exacerbated differences between the Allies themselves as to how to meet these challenges which could over time threaten the very existence of the alliance.

How to deal with these dangers, as well as the threat of armed attack in Europe, is the central problem for today and for tomorrow.

In these circumstances, the way the Atlantic nations, and their allies and associates around the world, should individually and collectively respond, both on the NATO front and on the wider global scene, whether through the alliance or outside it, will certainly call for some significant adaptations in outlook, policy and effort.

A review of the nature of the challenges and of the means of overcoming them must be undertaken by all concerned nations if the present predicaments are to be overcome. How this would affect the Atlantic alliance, which has so long been the centerpiece of security, is a natural place to start.

A combination of old dangers with new ones now faces the Atlantic allies with a much more complex, interwoven, multifaceted, sobering and potentially dangerous set of problems that ever before encountered by the alliance.

First, both the power and leadership of the United States relative to the USSR and to Western Europe alike has seriously declined over the past decade. To a certain extent, this was inevitable, as productivity expanded in both Western Europe and the Soviet Union. On the other hand, the retrenchment in American military investment and the psychological impact of Vietnam, coincident with the persistent growth of Soviet military power and activity, has not failed to have a pervasive political effect within the alliance in addition to its more obvious military implications.

Second, despite some improvements, the aggregate military strength of NATO has likewise seriously declined, in relation to the steady improvements in the Soviet military establishment.

Third, the threat to NATO is no longer focused entirely in Europe. Political instability in the Third World has provided the opportunity for Soviet as well as indigenous threats to Western access to raw materials, especially oil. Additional demands on the allies in terms not only of military resources, but in terms of political cost as well, result from this inescapable broadening of the NATO horizon.

Fourth, détente has succeeded in distracting a considerable segment of Western European opinion from the growth of Soviet strength in Europe. The very real benefits to Western Europe in terms of greater freedom of movement of peoples between East and West, favorable economic ties and the hope of continuing more stable political relations, lead the European allies to weigh with caution the potential costs of responding to aggressive Soviet conduct at home and abroad.

Fifth, There are increasingly divergent perceptions of the threat and, even more importantly, of how to go about responding to it, among the allies. The United States tends to emphasize the importance of military power; the other allies, to emphasize political and economic solutions.

NATO was born of a pattern of aggressive Soviet behavior. The consolidation of Eastern Europe after World War II to form a Soviet security buffer zone, the activities of Communist parties in Western Europe, Soviet pressure on Greece and Turkey, and the Soviet-supported invasion of South Korea, led the Atlantic allies to rearm and organize militarily within the NATO framework to fill the power vacuum which fairly invited Soviet domination of an unarmed Western Europe.

Today, that lesson of the deterrent value of military capability in keeping the peace has to be learned all over again. Not only in the Persian Gulf, but wherever relative power vacuums have existed, an invitation for aggressive adventurism is inherent in the situation. Military power can of course be used for armed conquest. Equally, it can be a decisive factor in the application of political, economic, and psychological influences and pressures short of armed force. The Soviets, in discussing their concept of "correlation of forces," expressly recognize that more than military power is involved in a total balance of forces but accord to military power the role of *primus inter pares.* Countervailing military power, then, is necessary to keep the peace.

But the military power of any one nation alone is far from sufficient to cope with the challenges involved. Collective security measures are the only possible answer; and collective security measures can not be maintained if there is not a fundamental political commonality of purpose between the nations participating in the effort. That common approach is now in danger of eroding because there is a growing lack of agreement on the nature of the problems and on what should be done about them. How to restore a common approach is of at least equal importance with the restoration of the balance of power.

The problem seems to lie in the paradox that the European allies look to American military power as essential to hold the Soviets in check, while at the same time they believe the United States tends to turn excessively to military solutions of international crises. This places the United States in a "damned if you do, damned if you don't" position, and has frequently led to American impatience with the reluctance of the European allies to follow its lead when it takes sides more definitely than the latter have considered prudent. Bridging this gap in approach is one of the main tasks for the Atlantic allies.

To find the resources to meet all these growing needs will be no simple task. The major industrialized nations face difficult problems stemming from the energy crisis, the rising cost of oil, social welfare programs that absorb the greater part of national budgets, and the general stagflation that is now prevalent in the world. It is therefore necessary to look anew at what is needed to keep the peace and the ways those needs can be met.

Allied cohesion and the political will to contribute, separately and jointly, to the common defense depend upon informed public understanding of the issues involved.

In political and psychological terms, Western civilization has evolved beyond a belief in the use of armed force as an instrument of policy; the Kremlin has not. In these circumstances public support, and especially youthful support, can be obtained only for more fundamental ends than the maintenance of a purely military alliance.

Despite all its troubles, the enduring strength of NATO has been, in the words of the preamble of the Treaty, the determination of its members "to safeguard the freedom, common heritage and civilization of their peoples."

The dignity and liberty of the individual, his freedom to worship in his own way, to choose his own form of government and to seek a better life for himself and his children are universal human values. They are common to all mankind, including peoples subject to totalitarian rule. These values represent not only the fundamental basis for allied cohesion but also its essential strength in the battle for men's minds everywhere. Understanding of the need to safeguard them can go far to strengthen the morale of our armed forces, willingness to serve in them, and political and parliamentary support for them.

In today's turbulent world, greater public realization is necessary that no nation by itself can assure its freedom and the basic values of its own people, but that common effort with other like-minded nations is essential to maintain them. Yet the younger generations today have no personal memories of *why* it was necessary to fight World War II, of the sacrifices and suffering which resulted from lack of adequate deterrence, or of the dedicated efforts in the early postwar years to build a better international order. The political will to safeguard freedom and the common heritages of the past has a tendency to decline from generation to generation unless it is constantly reinforced.

The economic, financial, and political problems which face the nations of the alliance are, by their very nature, long-term problems. So also are the problems of defense and of East–West relations; and the problems of defense buildup and economic health are inextricably intertwined. The predicaments affecting NATO which are analyzed in this study, and, to a considerable extent the proposals that are put forward, should be viewed in that light.

Major shifts in strategy, in approach, and in internal relations within an alliance take time to gestate and bear fruit. Their growth cannot be forced. But a beginning must be made. It is none too soon for the allies to start thinking together about these problems and together to seek the answers. The major predicaments for the Atlantic allies to consider are these:

1. The expanding threat outside the NATO area
2. Today's threat in Europe: the nonmilitary aspect
3. Today's threat in Europe: the military aspect
4. Is NATO strategy and force posture up to date?
5. Are NATO's reinforcement plans still realistic?
6. How can all these defense needs be met?
7. Internal and external problems of the alliance.

Only in jointly analyzing these problems can the nations concerned come to jointly recognize the course which must be steered and the obstacles that must be either avoided or overcome.

THE PREDICAMENTS

The Expanding Threat Outside the NATO Area

As the European allies withdrew from "East of Suez," and the United States shrank during the 1970s from its sense of global responsibility, the expanding Soviet military establishment has been used to extend the influence of the USSR into critical areas being vacated by the West. This same period has been one of change, unrest, and even turmoil throughout the Third World. The problem of economic development has been compounded by the uncertainties that have afflicted the economic and financial systems of the world and particularly by the impact of rising oil prices, while struggles for leadership and changes in the structure of societies have added to the kaleidoscopic nature of the situation.

The Soviets have taken advantage of this unsettled climate to build up their own military power and to enhance and employ the power and influence of their militant surrogates and dependents. From North Korea to Vietnam, from Afghanistan to South Yemen to Ethiopia and Libya, and across the Atlantic to Cuba, they are exploiting a chain of like-minded regimes for the reduction of Western influence abroad and the export of unrest and instability in the cause of Communist imperialism.

Under the guise of the doctrine calling for Soviet support for "wars of national liberation," the Soviets and their proxies have fostered subversion and insurrection in order to replace unfriendly regimes—particularly those friendly to the West—with those more supportive of their own hegemony. To this end the Soviet Union has stimulated and supported this coterie of dependents with massive military supply. Concurrently there have been continued Soviet attempts to undermine regimes friendly to the West, as well as to undermine the influence of the Western nations with the nations of the Third World, such as by massive Soviet propaganda attacks which have from time to time been surprisingly effective. Despite the fact that the Western industrialized nations provide virtually all the economic aid for the developing nations and the Soviet Union virtually none, Moscow's propaganda has somehow managed to portray the Western allies solely as militarist and imperialist, as the champions of oppressive political systems and the opponents of change and development.

Moscow finds these alternatives preferable to direct involvement in Third World countries. The use of Soviet military forces, whether to interfere with Western access directly or to intimidate a supplier nation from continued friendliness with the West is not the only available means. The less direct, more difficult to deal with and more likely to be encountered line of attack involves the exploitation of radical indigenous political forces to obtain the same results by influencing, intimidating, or even overthrowing governments too friendly with the West.

The extent to which Moscow is involved in this second type of threat will of course vary. Not every instance of anti-Western movements is the result of Soviet activity. But in all too many cases there is the pattern of militant groups trained, supplied, supported, and sustained by or on behalf of the USSR. They may want to serve their own purposes primarily, yet at the same time in fact they also become surrogates for Soviet action.

There are clear advantages for Moscow in this second course of action. The surrogate acts as a political insulator of Moscow, which can deny responsibility for the acts of its agent. Without becoming overtly an intervening power, the Soviet Union thus has a better chance of escaping opprobrium. The risks for the Soviets, militarily and politically, at home and abroad, are materially reduced, win, lose, or draw. It seems probable, therefore, that Moscow will continue to employ this approach as its preferred course of action. At the same time, as Afghanistan reminds us, the use of Soviet armed forces themselves for intimidation, or if necessary, actual interventions remains available as an option if the stakes are high enough and the risks not forbidding. It is against the background of local and regional political issues, just as much if not more than that of East-West confrontation, that the Western allies will have to learn to mount an effective political defense for themselves and the regimes friendly to them.

Whether the stakes are access to oil in the Persian Gulf, the quarantining of militant Communism in Central America, the continued independence of Thailand, or the freedom of the Malacca Straits in Southeast Asia, Western interests will require considerably more than military power to generate the sort of indigenous strength that can prevail over the continuing radical pressures that must be expected in the years to come. In the final analysis the Western nations must increasingly foster common interests between themselves and the developing nations of the Third World. Western support for progress in economic and social growth and justice is a necessary foundation for the sort of local resistance to militant expansionism that is so essential in the circumstances.

It is with regard to the diagnosis of particular Third World crises and the prescription of measures to improve the situation that differences between the allies are frequently to be found. The extent to which a situation of unrest and turmoil is fundamentally internal and indigenous in its roots and that to which it is attributable to outside interference is often difficult to decide. The degree of need for military support of regimes in power compared to the degree of need for pressing for social reform is often a place where reasonable men can differ. In these cases the United States and its allies have at times been at odds; and the European perception of an American tendency to favor military solutions has gained strength.

The threat of direct Soviet armed intervention outside the NATO area cannot be put aside even though it may be less probable than the indirect forms of intervention that have just been discussed. The Soviet invasion of Afghanistan underlines the fact that directly, through its own armed forces, as well as indirectly through surrogate nations, the Soviet Union has extended its reach to one of truly global proportions.

In addition to its impressive strategic nuclear power and its transformation of Warsaw Pact forces into a potentially offensive fighting force, it has, over the last decade, spread its power to most of the globe. The large, growing, and modern Soviet navy has increased its presence along the sea lanes of the Mediterranean, the South Atlantic, the Indian Ocean, and the Pacific to a far greater extent than required to protect Soviet maritime traffic to and from Siberia. Europe, the Middle East, most of the Indian Ocean, and the East Asian perimeter are all within range of the Soviet SS-20 and the Backfire bomber. The Soviet ability and readiness to project forces abroad, demonstrated in its military airlift support for Ethiopia and its invasion of Afghanistan, has not been without impact throughout the Third World and the industrialized nations.

No wonder, then, that the situation in the Middle East/Persian Gulf area, after Afghanistan, assumed crisis proportions. Nor is the fact that the Soviets may shortly become net importers of oil a reassuring thought. The Soviets now are in a position to threaten Western access to oil, whether in order to increase their own entree to this source of supply or as a means of bringing pressure to bear on the Western allies. The stakes for the West are truly vital. Middle East oil is indispensable for Japan's economy. It is only slightly less so for the European allies, and it is very important for the United States.

In concrete terms the issue is continued unimpeded access to oil, but there is a broader significance as well. Freedom of commerce and navigation are vital interests for the mercantile nations of the free world. Access to other strategically important natural resources (such as titanium, manganese, chromium, and cobalt, to name but a few) as well as oil, and even the very movement of goods and services of all sorts which makes up peacetime commerce may in turn become subject to more pressures than they have been in the past. The precedent in the case of access to oil is hence of major significance.

In the absence of a Western determination to protect access to oil, the situation could well invite further Soviet adventurism. A deterrent military presence in the area, supported by a reinforcement and resupply capability, is thus an inescapable requirement to protect the vital interests of the West. Although the Middle East is admittedly one of the least favorable locations for Western military operations—for example, time and distance factors, climate, and terrain—and needed base rights are hard to come by, the ability to deploy and sustain a consequential force in the area is all the more essential if Soviet adventurism is to be made obviously a costly experiment.

The Middle East is by no means the only trouble spot outside the NATO area, though as the longest neglected one, and the one affecting NATO nations most immediately, it has required priority treatment. In Northeast Asia, Soviet strength and that of its military dependent, North Korea, have also grown as United States strength has waned. The great bulk of Soviet ground force strength in Asia is deployed along the Chinese border. Soviet naval deployments, now include some eighty combat vessels and a like number of submarines, together with a growing air capability. Backfire bombers and one aircraft carrier now add a further dimension to their capability. The large Soviet-equipped forces of North Korea, sixth largest in the world, remain a constant threat across the DMZ, while Japanese-Soviet relations are at best uneasy. Any further significant reduction in American or allied capability in the area could invite instability and confrontation.

In southeast Asia Soviet support for militant Hanoi, and Soviet access to the naval and airbases of Vietnam, make the ASEAN nations, particularly neighboring Thailand, insecure and ill at ease.

The Western hemisphere encompasses considerable poverty and discontent as problems of economic development and social justice confront most governments. Social change, especially where long overdue, is being sought by violent means in many cases. This is a happy hunting ground for Soviet-inspired Cuban activity in support of local left-wing, radical, and militant forces, as exemplified by its activities in the Central American area. The struggle for stability is likely to be long and taxing and to place further demands on American political, economic, and military resources.

Deterrence, then, has perforce become a worldwide responsibility in response to worldwide threats. While clearly not every outbreak of hostilities in the world should be attributed to Moscow, nor is susceptible of solution by solely military means, a demonstrated capability and determination to respond to Communist assaults upon security, whether direct or indirect, should go far to let Moscow realize the costs of such ventures are going to be high.

Recognition of the necessity for defense against these problems outside the NATO area raises the issue of which nations should take the lead in responding to the challenge, and the mechanism that should be used to coordinate such activities.

It is necessary to reemphasize that the challenge outside the NATO area is not only military but political and economic as well. The defense must equally be at the political and economic levels as well as the military. In coping with the problems of the Third World, assistance in solving problems of economic health and social justice can be at least as, if not more, significant than assistance in developing the ability to resist armed insurrection.

The sine qua non for the success of any measures to improve security in the developing world is that they have indigenous roots and indigenous support. It cannot be repeated too often that viable solutions cannot be imposed, let alone maintained, by outside governments alone. The industrialized nations must learn to increase their respective abilities to work with and through people of the Third World and their governments. In the longer run security in the Middle East is going to depend on the nations of the Middle East themselves working together for collective security, with Western help and support, rather than primarily on Western military forces.

To these efforts, on this broad front, all of the industrialized nations can contribute to a more significant extent than in the past. At the same time, the greatest cost in terms of resources and responsibility has to do with regard to the military side of the equation. Here, the task has fallen primarily upon the United States, not only in the Far East and the Western Hemisphere, but now also in the Persian Gulf. At the same time, the concurrent presence of French and United Kingdom warships in the region and the supporting measures taken by Australia are most significant. Without this demonstration of common cause, the United States would all the more easily be identified in the minds of many as the sole concerned industrialized nation, materially weakening the Western position.

To organize to meet this complex global challenge requires flexible procedures that will involve all concerned nations, whether within or outside formal alliances. Consultation must insure that the views of all nations affected by the question at issue are thoroughly considered and that there will be no surprises between friends and allies. Certain nations, more willing and able to take active part in the necessary measures, will likely emerge as the leaders of one effort or another. But rather than to establish any formal directorate, with all the problems of exclusivity that would create, or to constitute some new body so broadly based as to be ineffective, it would seem best to use and expand existing multilateral and bilateral consultative procedures as needs be and set about getting the job done.

Today's Threat in Europe: The Nonmilitary Aspect

The Soviet-West German peace treaty and related international agreements, in effect recognizing Eastern Europe as a Soviet buffer area, went far to meet the overriding concern of the Soviet Union for its own security. Second only to that goal has been the Soviet objective of becoming the dominant power in Europe, to which it has since been free to turn.

The Soviet Union does not seek a war with the NATO nations. Both Soviet doctrine and Soviet opportunistic pragmatism call, rather, for a policy of political attack. Its policy is to obtain military superiority, first of all for defense, and then in order to exploit that superiority for political purpose. Frontal armed conquest of a strong opponent could be far more costly and inevitably involve more risks to the Soviet regime—already troubled with economic ills and the seeds of political dissent—than measures short of war. The prime mission of the Soviet armed forces, Soviet writers say, is to protect the system should a declining capitalist world be impelled to use force, and not to put the system at risk by moving against a strong adversary. Nor is Soviet policy likely to want to see Western Europe leveled in warfare. It would undoubtedly prefer to turn it into a docile, productive, and supportive buffer area. The most likely threat, then, is going to be in the political arena. The Soviets will try to combine fear of their military superiority with hope on the part of Western European nations for benefits from selective détente in order to support a major effort in order, over time, to split the Old World from the New.

The division of Europe from the United States as a requisite first step towards political domination of Europe may begin to seem to the Soviets to be attainable. Most likely, they will seek to decouple American nuclear power from Europe, to pare down American influence with its allies, to get the United States to accept Soviet inspired regional arms control arrangements in the name of Allied solidarity, and to play on the inevitable divergencies of European and American policy, especially in the Third World.

The Soviets have concurrently employed two principal strategies. One has been to foster the relaxation of tensions, not only with Western Europe, but originally with the United States as well. The other has been to sharply change the military balance through the major buildup of its own forces, and by seeking to weaken the NATO military position through arms control talks and by exhorting and threatening the European allies over the risks connected with nuclear modernization.

While following a policy of aggressive political and even military intervention in the Third World, in Western Europe the Soviet policy of "selective détente" has been marked by improved economic and political relations. Communist trade with European NATO countries is now about two-thirds of the current level of trade between the United States and its European partners. The hard currency supporting the Soviet share comes from Western credits. With Soviet exports to Western Europe focusing on the key areas of energy and raw materials, the politically important sectors of Western European banking and industry, as well as labor, are affected more than the bare figures indicate.

Western Europe is moreover important to the Soviets as much as an exporter of much needed technology as of finished goods. As far as the Federal Republic of Germany is concerned, not only economic gains but the improvement of contacts between the hostage peoples of East Germany and those of the Federal Republic has been a natural and preeminent motive in its support of détente. Finally, the replacement of bluster and threat with more proper Soviet conduct toward Western Europe, notwithstanding its pattern of conduct elsewhere, has influenced the political climate in the capitals of Europe and has gained support from a broad spectrum of society, including both left and right, that not unnaturally prefers détente to a return to the cold war with its apparently greater dangers and lesser benefits.

There are to be sure, some constraints on Soviet conduct in Europe as well. The Soviet economy is far from able to satisfy the desires of its peoples, and actions that would jeopardize the support it receives from East-West trade would be costly to it. But far more importantly, the chronically uncertain political situation in the countries of Eastern Europe must be a restraining factor in any estimate of the situation developed in Moscow. As the continuing unrest in Poland bears witness, the citizens of Eastern Europe are something less than sycophants of Moscow. What has been a security shield for the Soviets will not necessarily prove to be either a dependable source of military strength or a dependable line of communications in the event of Soviet attack.

All of these factors argue against the probability of an armed attack against Western Europe. Far more likely is the use of Soviet military strength to back up pressure, whether tacit or overt, directed toward influencing and dominating Western European policies to support Moscow's ends.

If the challenge of the future is likely to be in the political area rather than on the battlefield, does it follow that it is no longer necessary to maintain a balance of military power between NATO and the Warsaw Pact?

The answer is self-evident. If the balance of power were allowed to become so askew that resistance to military pressure, let alone armed attack, would be impossible, the extension of Soviet hegemony would be all too easy. The Western values of democracy and respect for human rights and individual freedom would be, in all too short a time, seriously menaced, or indeed, snuffed out.

In addition, while the risk of armed attack is small, there are still some genuine risks of hostilities that must be taken into account, including:

- The danger that the Soviets would be more likely to risk military adventures in the Middle East or elsewhere which could spread to Europe should they come to feel that the balance of power in Europe was strongly in their favor.
- The ever present possibility that, as in the past in Czechoslovakia and Hungary, or, more recently, in Poland, the Soviets would use armed force under the Brezhnev Doctrine to preserve their hegemony in Eastern Europe—or in connection with Berlin—with ensuing events leading, intentionally or unintentionally, to armed confrontation between NATO and the Warsaw Pact.
- The possibility that the Kremlin might feel forced to evoke the Russian people's love for "Mother Russia" in a military adventure should the inherent human hope for freedom spread from Poland and other Eastern European countries to influence the Soviet Union itself.
- The possibility that hostilities along the Mediterranean littoral, whether instigated by Libyan militancy, Arab-Israeli confrontation, or otherwise, could subsequently lead to an East-West confrontation.
- The danger that a continually growing imbalance of military strength could, over time and accompanied by a resulting political weakness in the West, lead the Soviets to risk a surprise attack in anticipation of early collapse and surrender by the West.

However, it cannot be gainsaid that these very favors have led to a most troublesome reaction which is increasingly being reflected in vocal segments of European opinion. Demonstrated primarily by opposition to nuclear arming of the West (and begging the question of the much greater nuclear arming of the East) and by the erroneous concept that there is little that smaller countries can do to affect the balance of power, it has developed into a neodefeatist school of thought that is reminiscent of the "Better Red than dead" slogan of some years back. While still a minority, any movement that draws strength from such diverse sectors as church and youth groups and

Labor and Socialist elements in a number of Western European nations presents real cause for concern.

Building up Western defenses is therefore only half the story. The other half is going to be in the field of political and diplomatic action to seek to cope with the political maneuvers of the Soviet Union which are short of open threat or force. This is in many ways a more difficult task than that of building a credible defense posture. It means dealing with the minds of men — allies, friendly powers, and opponents alike — and doing so within the tradition of freedom of thought that is the hallmark of Western civilization. It means developing a consensus on what should be done with regard not only to the global threat, but even more importantly, with regard to the continuing ever present political aspects of the threat to NATO.

It means demonstrating continuing willingness to enter into negotiation for meaningful arms control agreements which will restore or preserve stability and the overall balance of power. For the European allies commitment to the pursuit of arms control measures is an indispensable concommitant to measures of defense. The United States must be responsive to this political requirement, as, for example, with regard to the dual track policy on the theater nuclear force decision of December 1979, which linked arms control negotiations with the decision to proceed with the long range tactical nuclear force. The current French proposal for a first-phase conference on disarmament in Europe, focused on confidence building measures, exemplifies the opportunities that can be taken to preserve and enhance security.

It means relying more on measures short of the use of armed force, reliance on the good offices of other nations, the employment of political, economic, and financial leverage, and the development of strong pressures from the entire international community. It will encompass employing bilateral diplomatic channels, multilateral organizations, and private channels alike. It will mean increasing the public consciousness of the average citizen of the West of the risks and dangers of the coming period — a task made doubly difficult as the generation that has experienced the threat of Soviet power yields to a successor generation that has not known these dangers at first hand. It means doing all this in a way that strengthens the political support for the alliance both internally within each nation and externally between them all. And above all, it will mean devoting conscious effort on the part of all the nations of the alliance to maintain their unity despite the centrifugal pressures and temptations that must be planned for and resisted. Without that unity, there will be, as Senator Arthur Vandenberg warned, surrender on the installment plan.

It means involving non-NATO countries in cooperative efforts to deal with security problems in those regions. It will involve reversing the recent trends toward lower informational activities such as Radio Free Europe and the Voice of America while Soviet broadcasts multiply dramatically. It will require better and more timely intelligence better disseminated on the part of the free nations.

An uneasy peace is likely to be what the Western nations will have to acclimatize themselves to for a long time to come. Sustaining vigilance and common cause over an extended period which may never provide the stimulus that comes from the imminent fear of war will not be easy, but it is likely to be what is required.

Today's Threat in Europe: The Military Aspect

It is in the European theater that the Soviets have consistently deployed their best-equipped, most ready ground and air forces with 30 combat ready divisions and 1,700 aircraft deployed in Eastern Europe. Backed up by the less effective but numerous second echelon forces in the neighboring USSR, their presence inescapably requires a high level of countervailing forces in Western Europe.

The Soviet Union has pursued a program of steady quantitative as well as qualitative improvement over the past decade. New generations of tanks, guns, armored fighting vehicles, and aircraft, along with an impressive chemical warfare capability, have transformed the Soviet military establishment into a first-class fighting force with improved combat power and sustaining logistical support. Its strength and composition far exceeds what might be needed either for defense or to preserve internal security in the Eastern European countries. It has now clearly attained an offensive posture as well, increasing the need for readiness among NATO forces.

While the NATO nations have continued to modernize the equipment of their forces, they have failed to keep pace with the gains of the Warsaw Pact. The NATO long-term defense plan of 1977, intended to remedy some of these deficiencies, has not been implemented with the sense of urgency and of additional priority needs that it should command. It is at best a shifting of priority of defense spending, being implemented by deferring other high-priority needs.

NATO readiness has always needed to be at a high standard to overcome certain inherent disadvantages to which the alliance has been subject, some inescapably, others at high cost. NATO, as a purely defensive alliance, gives the USSR the initiative that comes with attack. The Soviets also have the advantage of internal lines of communication, uncertain though they may be, while the alliance depends on sea transport for much of its sustenance.

Then, too, there is the fact that the peacetime deployments of too many of NATO's troops are not garrisoned at their optimum battle stations because of the high economic and political costs of relocation. And there is the political requirement of "forward defense," that is, that the land of one ally cannot be sacrificed to protect the others, which excludes the possibility of trading space for time. Though relocating cantonment areas and dropping the forward defense strategy might be desirable from the military point of view, the cost of the disruptive political strains that would accompany such actions would outweigh the military advantages that would accrue.

The most serious deficiencies are in those readiness measures needed to let NATO's deployed forces fight successfully against large scale attack at the conventional level, particularly in the central region. Despite some steps to improve the situation, the deficiencies still include:

- Lack of defense against and response to chemical warfare, both still inadequate in the allied forces in comparison to the threat;
- Deficiencies in operational reserves;
- Inadequate numbers of tanks and antitank weapons to counter the Soviet armored superiority.
- Shortage of ammunition of all sorts, currently inadequate for sustained combat conditions of more than a few days, and in many cases poorly located and protected;
- Shortage of other war reserve materiel stored in the theater to replace materiel consumed or destroyed in combat;

- Inadequate air-defense measures, particularly surface to air missiles, compared with excellent Soviet capability in this area, presaging serious NATO losses early in any air battle;
- Lack of survivable war headquarters and interface of allied and national communication networks;
- Insufficient allied war headquarters manned in peacetime;
- Lack of sufficient exercises in which national forces are placed under international command;
- Shortages in infrastructure of all kinds, national and international;
- Lack of naval forces needed to insure the arrival of reinforcements in Europe;
- Deficiencies in trained manpower in ready and reserve units;
- Lagging electronic warfare measures to counter the impressive Soviet capability in this field; and
- Inadequate training of personnel and maintenance of equipment.

Moreover, ministerial level government officials do not sufficiently appreciate how national legal and procedural obstacles could hinder NATO from quickly adopting an alert posture in time of crisis. It must be remembered that, except for air defense forces, most NATO forces remain under national command in peacetime, even including, for example, the intelligence and warning activities of the reconnaissance forces in the central region. Each of these democratic nations must move within its legal procedures to authorize essential preparatory measures that should be taken in a period of strategic warning—a slow enough process even if well prepared for in advance, which is not now the case. Governments are themselves not yet prepared to overcome concerns that moving to a state of preparedness in response to a crisis situation would increase tensions, a theme long exploited by Soviet propaganda.

These problem areas alone could limit the duration of the conventional phase of hostilities to a matter of only days, or at best, a few weeks.

Because of the greater Soviet military pressure in NATO's central region, the special problems of the northern and southern regions are all too often overlooked. But these regions stand astride the exit routes from the Baltic, White, and Black Seas, affecting access to the

Atlantic, Mediterranean, and the Middle East, together with the Persian Gulf. Turkey is the land bridge to the Middle East and Eurasia, standing between the USSR and the Middle East. The security of these areas affects the security of all NATO as well, and is equally important in Soviet strategy. The Soviets maintain continuous political pressure on both regions, by such measures as the movement of elements of their armed forces, pressing the Svalbard issue, and, most recently, attacking prestocking in Norway and reviving the proposal for a nuclear free Scandinavia, and through economic and political overtures to Turkey. They remain quite ready to exploit internal difficulties in any flank country which might be used to weaken its ties to NATO.

The problems of these two regions are compounded by the longer distances which separate them from the rest of the alliance. There are political problems in the northern region limiting peacetime foreign military presence, while in the southern region the heritage of the recent Greece–Turkey strife requires extensive rebuilding of the indigenous economic and military power base. Particularly as the additional military requirements for the Middle East are being considered, it is essential that the ability of the alliance to support the countries of the flanks remain significantly high.

In time of crisis, smaller or more troubled nations need the assured support of their allies. The presence of the Sixth Fleet in the Mediterranean and the NATO roles of United States military units at installations in the southern region have signaled that an attack in that area will not involve local forces alone. In the north the preparatory prestocking for allied reinforcements should convey the same message; and combined training exercises can further demonstrate allied unity. The participation of various nations has been deemed politically important, as evidencing broadly based identity of interest on the part of the allies in the defense use of resources. Air defense and communications still need improvement. The overall very powerful political–military pressure in the Nordic area operates to make defense all the more difficult. In the south the return of Greece to the NATO integrated military fold is most welcome news, but Turkey's extreme internal problems and the unfortunate hiatus in U.S. arms supply have created continuing political and military problems of no small dimensions.

One of the better ways to speedily show many flags, and so to demonstrate allied resolve in a time of crisis, is the use of combined forces, such as elements from the Standing Naval Force Atlantic (STANAVFORLANT), and the Allied Command Europe Mobile Force, which can promptly be injected into an endangered locality. The ability to deploy such multilateral forces in response to pressures to position them on both flanks simultaneously could be important. They would underline the risks for the Soviets that attack on the flanks would involve real risks of widening of the theaters of war.

The ability to resupply NATO's first-line troops takes on added importance in light of these problems of readiness. This immediately brings to mind issues of the availability of manpower, equipment, and transport. European reservists are mainly dedicated to fill existing units and to be individual replacements rather than to constitute replacement units, mainly because of the lack of the additional unit equipment needed for the latter role. American reserves are notoriously short of personnel, particularly for the individual ready reserve. Both are deficient in training. The status of equipment for resupply is less than reassuring. Stockpiles of finished equipment are not available; nor in many cases are there adequate stockpiles of raw materials to permit production. Industrial mobilization is not organized or planned for in any adequate way on either side of the Atlantic, to provide a "surge" capability in time of crisis or warfare.

The transportation system must cover not only military but civilian needs as well. Heavy equipment and bulk supplies will have to be moved by sea and perhaps under combat conditions. Many European allies rely on United States replenishment of American-made inventory items. It has been computed that NATO will need some twelve to twenty-four dispositions of 70 merchant ships each at sea at any one time. The weakness of the American merchant marine has to some considerable extent been countered by the agreement of the European allies to make some 600 merchant ships available in time of emergency, but problems in assuring needed civil airlift support have not yet been resolved.

The task of receiving material in the theater and moving it up to the combat troops, a national responsibility, still is inadequately prepared for and loosely organized, despite reason to hope for progress in current negotiations for increased host nation support. The security of ports is threatened by the long-range Soviet bomber and fighter bomber aircraft as well as missiles. The opportunities for mining of ports and for sabotage of supply are self-apparent. The capabilities of landing "over the beach" or by lighters are now being explored, but civilian ships' crews need to be made familiar with these measures. The nettlesome problem of refugees will further complicate the movement of men and equipment.

The ability to reinforce NATO's front-line strength is one of the most significant aspects of NATO's ability to mount a credible deterrent. The ready, forward-deployed forces in Europe do not provide any forces for a strategic reserve for the central region, which is sorely needed to give credibility to the NATO defense. It is now doubly important in light of the improved readiness and offensive posture of the Soviet forces. To strengthen the defense and to provide a strategic reserve, the United States is in the process of pre-positioning heavy equipment for four divisions in Europe, with equipment for two additional divisions planned to follow. This will permit personnel to be deployed by air, cutting deployment time to about 10-15 days after receipt of strategic warning and national decisions in response thereto. The ability thus to strengthen NATO defenses, in such a time of emergency—indeed, to provide a reserve pending the arrival of U.S. reinforcements—could be of the greatest political as well as military importance. It would demonstrate both the strength and the resolve of the allies, notwithstanding the massive Soviet propaganda barrage about "tightening of tensions" which must be expected at such a time.

To what extent NATO can and should continue to rely on these plans for reinforcement from the United States, particularly in light of the additional responsibilities the United States is now assuming in response to the global threat, is the subject of more detailed consideration in a subsequent section of this policy paper.

But even under the more favorable assumptions of present plans, NATO cannot have high confidence of maintaining a forward defense without early resort to nuclear weapons. The heavily increased risks for NATO, stemming from the loss of that nuclear superiority on which it has long relied, and what it implies for the credibility of the deterrent, are dealt with in the next section.

Is NATO Nuclear Strategy and Force Posture Up To Date?

The central strategic nuclear capability of the United States remains an indispensable factor in the balance of power. Better than expected performance and earlier than expected deployment of the fourth generation of Soviet ICBMs, concurrently with the delay of U.S. strategic modernization programs, has nevertheless given the USSR general parity, if not in fact impending superiority. That trend is now in the process of being changed. Concern over the continuing survivability of the United States land-based missiles in light of growing Soviet missile strength and of the continuing capability of the other legs of the triad is now leading to measures to redress such deficiencies. Inferiority at the strategic nuclear level would drastically affect the confidence and political fiber of all the allies. Although the superiority in strategic nuclear weapons the West enjoyed in the past may no longer be attainable, inferiority simply cannot be accepted, as the American people have recently made clear. Maintaining general strategic parity is hence a matter of such imperative priority, that, in the ensuing discussion it is assumed that it will be maintained by such short-term and long-term measures as may be necessary.

At the theater nuclear level, the deployment against Europe of the Soviet SS-20 mobile IRBM represented a substantial improvement in the Soviet nuclear arsenal, giving the Soviets a prompt counterforce capability against all of Western Europe, a capability not matched by NATO. Together with the Backfire bomber and the new generation of long-range fighter bomber aircraft, this survivable long-range threat to Europe has tipped the theater balance heavily in favor of the Warsaw Pact.

The relatively small long range theater nuclear force (LRTNF) is at once a first step toward restoring the theater nuclear balance and a first step toward meaningful arms limitation negotiations on theater nuclear weapons. The LRTNF may serve as reassurance to those European allies who were concerned at the possible decoupling of the strategic deterrent. It is, however, of more than political significance in that it can provide an ability to interdict the reinforcement of Soviet troops in Eastern Europe and will bring Soviet territory within range of weapons based in NATO. But while the LRTNF should make Soviet planners less confident of the success of large scale attack, the fact remains that resort to nuclear weapons by the allies would be counterproductive should the Soviets respond with their full theater nuclear power.

The loss of the clear strategic and tactical nuclear superiority which marked the earlier years of NATO's history is obviously a matter for serious consideration. The extent to which it requires change or adaptation of NATO strategy and force posture is considered below.

NATO strategy has long been based on a combination of conventional and nuclear weapons. What the proper mix between these two elements should be has been a matter of equally long debate within the alliance. Some have favored a relatively low level of conventional defense. The arguments in favor of that position include defense budget limitations, concern that the ability to fight a sustained campaign on the conventional level would decouple the deterrent effect of nuclear weapons, worry lest Europe again be devastated by a great conventional war as it was in the 1940s; and, rightly or wrongly, a feeling that the Soviets are not going to attack the West even if there is only a low level of allied resistance. These arguments are found mainly, but not solely, on the European side of the Atlantic.

Others have urged a higher level of conventional defense. They have believed that a conventional defense which could only hold for a matter of a short time would be no better than a tripwire leading to nuclear war at an unnecessarily early stage. They have also been concerned that too thin a conventional defense would weaken the credibility of NATO's deterrent and defensive capability, as it would weaken the Soviet impression of the readiness of the NATO countries to move from deterrence to actual war if attacked on a large scale.

The end result of this debate has been a compromise in doctrine and a conventional force and equipment level that has so far failed to meet even the levels that doctrine requires. NATO's conventional forces are supposed to be able to sustain a forward defense sufficient to inflict serious losses on the aggressor and convince him of the risks of continuing his aggression. As we have seen, NATO's forces can in fact sustain such an effort for all too short a period of time. But then NATO has been able in the past to rely on the use of tactical nuclear weapons to deter and defend to the extent that the conventional capability has been insufficient. Western nuclear superiority in the past supported the assumption that escalation to nuclear warfare would place control of the situation in NATO's hands.

What has occurred to change the picture is the loss of that fairly clear NATO superiority in both strategic and tactical (or theater) nuclear weapons, which had given the Atlantic alliance escalatory control. Such clear NATO superiority is not likely to be seen again. The central strategic nuclear balance is likely at best to be one of general parity over any extended period of time. Meanwhile the Soviets have now gained superiority in the theater nuclear realm with its new generation of weapons, particularly the mobile, survivable

SS-20. In such circumstances, resort to tactical nuclear weapons by the allies would be likely to make them worse off, as not only the national infrastructure and populations of Western Europe, but also NATO's military forces themselves would be vulnerable to Soviet nuclear counterattack. To the extent that control of the situation through escalation to tactical or theater nuclear war exists, it can be said to now rest with the Warsaw Pact rather than the West.

This new situation calls for a thorough, new stocktaking. Its implications for the NATO deterrent concept are obviously significant. The first question is whether nuclear weapons are now of no significance other than to create a nuclear stand-off, leaving the conventional balance of forces to be the sole arbiter in the event of armed conflict. If that were to be the case, NATO might just as well choose to accede to the Soviet proposal for a "no nuclear first-use" treaty.

Deterrence depends on confronting an adversary with sufficient strength to make him doubt that armed attack could be profitable. Obviously the more doubt of success, the better; the essential task is to deny him a reasonable prospect of success. In this context, nuclear weapons, as long as they continue to exist, may well continue to contribute to the deterrent by making the situation too ambiguous for any would-be aggressor. The shift in escalatory control surely means that the Soviet Union is no longer faced with as high a degree of probability as before that the West would employ nuclear weapons if it were at the point of losing a war. The question then is whether there is now absolutely no chance that the West would resort to first use, or whether the Soviets would have to entertain some doubt whether nuclear weapons might yet be used.

As long as nuclear weapons exist and general strategic parity prevails, the risk of allied first use of nuclear weapons in response to armed attack by the Warsaw Pact, particularly in times of confusion and anxiety or as a measure of last resort, simply cannot be excluded from the thinking of the Soviet planners. The loss of escalatory control appreciably reduces the readiness with which NATO would be prepared to initiate use of tactical or theater nuclear weapons; at the same time, it does not go so far as to necessarily reduce that chance to zero. If large-scale Soviet conventional attack were met with a stubborn and sustained conventional defense but nevertheless was at the point of prevailing, the Soviets could not but be uncertain that such a resolute enemy would not turn to nuclear weapons as a measure of last resort. In such circumstances, with the conquest of a viable Western Europe no longer possible and with the survival of the Soviet regime and homeland at risk, the Soviet planners simply would not have the ability to guarantee a reasonable chance of success.

It remains to consider what the effect would be on Soviet appraisals should NATO's conventional capacity to resist an attack remain markedly more limited, for example, a matter of days or at best a few weeks, as is now the case. The Soviets might well draw quite different conclusions from such a demonstration of lack of purpose on the part of the alliance. They might base their estimates of the situation on European fears that the Americans were decoupling nuclear weapons because of fear of Soviet nuclear power, as well as on American reluctance to take the risks of strategic nuclear war on behalf of European allies which would not undertake a genuine conventional defense of their own lands. In the absence of NATO preparations for a more serious conventional level defense than in the past, they might then well be led to conclude that, if push came to shove, NATO would not really fight at all, with either conventional or nuclear weapons.

In other words the conventional balance has assumed far greater deterrent significance now than it did before, when NATO had clear escalatory control. A more genuine conventional capability is essential to make NATO's strategy of deterrence and defense responsive to today's needs. It must be bolstered significantly if the Soviets are still to think the alliance is serious. It must equally be bolstered to assure continuing confidence within the alliance that commitments and obligations are being shared equally and that the job of deterrence and defense is being successfully performed.

Some observers believe that a new strategic doctrine is called for, one which pointedly rejects early reliance on nuclear weapons and clearly demonstrates the willingness of the European allies to defend their homelands no matter the cost. But this would inevitably lead to a debate on nuclear strategic doctrine, which could be divisive, creating domestic political problems for some allies and sharpening divisions within the alliance. In any event it would not be necessary so long as a consensus developed simply reflecting a new sense of urgency about building up the conventional element of NATO's forces — a recognition that NATO's conventional strength, deficient even before the loss of nuclear superiority, is now on the way to being dangerously deficient.

Are NATO's Reinforcement Plans Still Realistic?

At the December 1980 ministerial meeting of NATO, there was agreement "to prepare against the eventuality of a diversion of NATO-allocated forces the United States and other countries might be compelled to make in order to safeguard the vital interests of weaker nations outside the North Atlantic Treaty area." The ministers then "recognized that the developing situation would entail a suitable division of labor within NATO." This agreement in principle remains to be fleshed out.

It is instructive to review how a major part of the reinforcement of the alliance's defense effort falls on the United States. At the risk of oversimplification, it can be said that there seems to have been a basic assumption that all-out war would certainly involve the NATO theater, and that that theater would have overriding priority over any other. In light thereof, all American units not otherwise committed, both active and reserve, have long been allocated to the reinforcement of NATO. Those premises now no longer fully apply, and hence it is necessary to consider how NATO's reinforcement needs can be met under the new circumstances of conflicting requirements.

Now it appears possible that some of these active duty American forces might have to be transferred to the Middle East should hostilities occur there at the same time as a NATO crisis. In addition, U.S. military air and sea lift as well as covering naval and air forces could be heavily involved in that theater just when needed to reinforce Europe. The all too few specialized service support units needed to receive, handle, and maintain equipment once arrived in theater might all be already involved in the Middle East. There could also be some derogation of projected NATO deployments, such as intelligence collection aircraft and even combat units, as the needs in other theaters developed. The heavy shortages now existing in reserve units and in individual ready reserve personnel, if not overcome, could further jeopardize the reinforcement of the European theater. And should there be hostilities, the safe and prompt arrival of reinforcements in Europe, whether from the United States or the Middle East, would become increasingly uncertain.

The importance of being able to add to NATO's ready forces promptly on receipt of early warning has already been noted. Common prudence requires preparation now against the contingency that some of these reinforcements, most probably those units whose equipment had not been pre-positioned in Europe, would not be available within the early time frame required.

It is not feasible to look to the United States in these circumstances, given the additional burden it is already assuming in the Middle East on behalf of the alliance. Moreover it must be assumed that available active U.S. ground forces would be already committed, whether to NATO or elsewhere, and the readying of reserve units from the United States would involve too great an element of time to be a useful measure.

One alternative would be to plan for the withdrawal of two or three divisions of U.S. ready forces from NATO Europe for deployment in Southwest Asia when and if needed. In terms of time and distance, Europe is much nearer to the oil fields of the Middle East than is the United States. At the same time, the political and practical problems of overflight arrangements, coupled with the political and military problems of such a force withdrawal from NATO should the crisis threaten the European front as well as access to the Middle East, make this alternative offer less than an adequately certain and effective basis for planning for either the NATO or the Middle Eastern theaters.

On the other hand, a further alternative strongly commends itself because it would improve, rather than lower, NATO defenses. The European allies have large numbers of reserves, but in many cases they are mainly intended to be individual replacements, and lack only training and the organizational equipment that would let them be quickly available for employment. To the extent that these reserves could be manned and equipped by the European allies, the alliance would be able to rely on a more assured, more immediate and less costly means for rapid mobilization and reinforcement by M + 10. Though some increased costs for training of the reserve personnel would be involved, the greater personnel expense of continuing peacetime active duty pay would not need to be incurred. In this time of need, this obvious resource should be fully exploited to meet the need for the "suitable division of labor" the North Atlantic Council had in mind.

In addition, at least two other steps are needed to compensate for the effect of Middle East emergencies on NATO's ability. The shortage of airlift in the event of concurrent crises in two theaters would affect the ability of the United States to provide air transport for the divisions which would remain allocated to NATO for M + 10 deployment. In light thereof, plans for alliance-wide civil air transport support should be reviewed and updated to insure prompt and timely arrival of reinforcements upon receipt of early warning. A most essential additional measure would be the expansion of host nation support services to ensure the reception and forward movement within the theater of reinforcement and resupply shipments, a subject now under negotiation. The combat support and combat service support functions which are performed by the all too few U.S. specialized units should be transferred to European civilian and reserve personnel, lest the U.S. units be already involved in another theater. Contingency planning for other compensatory measures may well become desirable as the situation develops.

Once the alliance deals with the immediate problem of compensating for the impact of Middle East requirements, it can turn to exploring means for meeting the growing need for improved conventional capability. The advantages that would accrue from developing the potential of reserves already in Europe are so significant as to suggest that this method of procedure also be increasingly relied on by the alliance in the longer term future. Organizing the trained reserves in Europe into additional combat units which would be speedily mobilized and deployed in time of crisis could provide the most efficient and effective solution to that need.

How Can All These Defense Needs Be Met?

From almost every point of view, the resources of the non-Communist world far outweigh those of the Communist world: population, gross national product (GNP), productivity, technological capability, demography, and that indispensable stimulus, freedom of thought.

Indeed, NATO Europe alone has two-thirds the population and more gross national product than the whole Warsaw Pact. If the economic strength of Japan is added to that of NATO, the superior strength of the industrialized nations of the West becomes even more apparent.

With regard to certain strategic natural resources, however, the allied nations are at some marked disadvantage. Indeed, it is surprising that the significance of those materials for a "surge" capability in military production has received so little attention. Minerals and metals such as titanium, manganese, chromium, cobalt, and aluminum must be imported by all the Western industrialized nations, while the Warsaw Pact countries can depend on their own resources for these commodities to a very great extent. Energy resources—particularly oil—are of critical importance to the Western nations. The lack of adequate stockpiling of such items has long been the subject of study within the alliance, but, with few exceptions, of inadequate action. But with these exceptions, it is clear that the allies have the resources in money and men to cope with any challenge of which the Warsaw Pact is capable.

The simple fact is that, while social programs have burgeoned, proportionately larger resources simply have not been made available for defense by the West, while the Warsaw Pact has given defense expenditures an overriding priority at the expense of social programs. At the same time, important military manpower problems have emerged which have been detrimental to the West. Terms of service have been reduced dangerously, and, in the case of the United States, problems of recruitment and retention of skilled and unskilled personnel have affected both the quality and quantity of what should be a proficient and fully manned force.

The fact that NATO has lost ground to the Warsaw Pact over the past decade is thus not due any more to the greater Soviet effort than it is to the inadequate Western effort. The sense of urgency with respect to defense measures has been lacking. Public support for defense clearly was flagging throughout the 1970s. Among the NATO nations, U.S. defense expenditures for NATO decreased in real terms over the decade. The expenditures of some of the European allies increased. But the net result has been a broadening of the number of deficient partners and a slowing down of the total defense effort. And even with the drop in the U.S. defense expenditures, the United States spent twice as much on a per capita basis as the European allies.

At the same time the economies of many of the continental allies blossomed. France, Germany, the Benelux nations, and Norway all approached or bettered the same level of GNP per capita as the United States. Yet the ratio of defense costs of these nations to those of the United States remained about 5 to 3, or at best, 5 to 3½. The United States is currently on the verge of undertaking major new defense programs in an effort to rectify the present imbalance. This will concurrently require painful reductions in other sectors of the budget. It is part of a larger effort to strengthen the economy as well as the security of the nation. At the time of this writing, it is too early to say with certainty how much defense spending will actually be increased, but indications are clear that, with a Reagan program averaging $35 billion increase per year for five years, there will be significant and sustained increase.

The result of these increases could bring the spending of the United States for defense close to the ratio of 2½ to 1 in relation to European allies whose GNP per capita is at least equal to that of the United States. Should European defense expenditures fail to respond to this sort of American leadership, the popular reaction in the United States can be easily imagined.

In terms of total resources, there is no question but that the United States and its European allies can increase their defense budgets substantially. The rate at which this can be done without adversely affecting the economy, let alone raising difficult political problems prejudicial to the common defense, is the practical question facing every government in the alliance.

Leaders in the NATO countries naturally want to ensure that increasing defense expenditures will not result in political defeat at the hands of opponents who have opposed NATO over the years. Equally, they wish, in time of inflation, stagnant growth, exorbitant energy costs, and economic and financial uncertainty, to preserve the health of their economies, so essential for political stability.

In fairness it must be pointed out that the picture is by no means all one-sided. One can take comfort from the return of Greece to NATO's fold; from Turkey's renewed pledge of constancy, from France's increase of 18 percent in funds for defense, and from Italy's attitude on theater nuclear weapons. Over the past decade, German expenditures for defense have increased more regularly than American. The European allies provide by far the greater share—over three-quarters—of the ground and air forces for the defense of Europe. They provide proportionately more than the United States in terms of aid to less favored nations, an expense in the shared interest of security in the Third World. Germany has been a leader in economic aid to Pakistan, Poland, and above all, Turkey.

Most of the allies rely on conscription, rather than the more expensive all-volunteer forces employed by the United States. Indeed, some European leaders have said that Americans cannot be considered serious about their armed forces until they go to conscription. And, most important, the average European citizen, himself subject to compulsory military service, views the American system as a shirking of equal responsibility and hardship on the part of the United States.

Even so, there is a strong sector of American opinion that holds that greater burden sharing on the part of the allies should be a prerequisite to further NATO defense contributions by the United States. Some even go so far as to argue that increases in American defense budgets should be matched dollar for dollar by countries whose GNP per capita matches our own.

Some believe that the United States should withdraw two or more of the U.S. divisions now stationed in Europe, leaving it to the European nations to replace them. The Atlantic Council's working group is emphatically opposed to reducing the American forces in NATO. In addition to being military folly at a time when NATO forces should be strengthened rather than weakened, it would send the wrong political signal to friend and foe alike. It would undermine European confidence in America and generate strong pressures for European accommodation with the USSR. Weakening America's role could beget further weakness in Europe; conversely, staunchness on the part of the United States should beget a greater degree of staunchness throughout the alliance.

Nevertheless, it must be bluntly cautioned that while the proposal for such a withdrawal from Europe is not the prevalent view at this time, American public opinion could very well turn in that direction if there was either a general belief that there was no longer a fair division of labor within the alliance, or that the allies were seriously at cross purposes in their foreign and defense policies.

The question of resources cannot be left without discussing the problem of burden sharing by nations outside NATO. With the per capita GNP of Japan approaching that of the nations of Western Europe and the United States, the share of Japan's contribution to worldwide peace and stability is inappropriately small. Even under the concept of "comprehensive security," whereby contributions to local and regional stability through nonmilitary measures would be included, it remains totally inadequate. Benefiting as Japan does from American security efforts in the Middle East and in Northeast Asia, as well as from the British and French naval presence off the Persian Gulf, it should not only bear an increased share of its self-defense but contribute far more heavily and directly to the expenses being incurred to its benefit. Other allies in the Pacific, such as Australia and New Zealand, can also further contribute to the common effort.

The industrialized nations of the free world are profligate with the defense resources that are available to them. The NATO allies, however, have all the disadvantages of a system wherein each nation is responsible for the supply of its own forces. The inevitable waste of duplication, with lack of interoperability of all too many kinds of equipment, has ensued, and, incredible though it may seem, has defied solution for over two decades. The cause has been a combination of selfish motives, including protection of national industry and labor, the "NIH" syndrome—whereby funds are refused for equipment "not invented here"—and just plain national pride. In all too many cases American units and bases in Europe cannot support or repair the equipment of other allies which might need their help; and the same is true in reverse. Mutual support of troops deployed on each other's flanks becomes questionable in these circumstances. Economically, the result is deplorable; it could be tragic operationally.

But on both sides of the Atlantic there has been a lack of true dedication to a genuine two-way street for defense procurement. Past experiences have been unsatisfactory, particularly due to the attitude of the United States; and future plans unfortunately are based on that track record rather than on trying to achieve greater efficiency in the employment of allied resources. Unless greater efficiency in the employment of resources is achieved, the alliance will continue to get less for each defense dollar than it should. Measures of rationalization, standardization, and interoperability (RSI) such as licensing and coproduction have been useful. But even in their totality, and if given full support, they could not provide the sort of savings and effectiveness that could come from more intelligent use of the research, development and procurement measures of the respective nations to meet the military requirements of the allied headquarters. Until national parliaments are willing to put protectionism aside in favor of the common good, however, maximum effective use of resources cannot be attained. The increasingly imperative need to bolster NATO's conventional posture should help to break through old prejudices which can but ill be afforded today. In the meantime RSI measures should be backed even more strongly than in the past, with the United States for once taking the lead in action as well as words.

Recapitualation: External and Internal Problems of the Alliance

It has already been pointed out that the most probable danger to the alliance and the highest priority Soviet policy goal concerning the alliance is the splitting of Europe and the United States. The Soviets can be expected to exploit fully such differences as exist, playing on them to exacerbate their impact. And such differences stemming from divergent interests and perceptions not only exist, but to a considerable extent are inherent in the transatlantic relationship. Moreover, we have seen how facing up to the need to update NATO defenses is bound to create additional internal problems for the allies.

To recapitulate, the major issues that are likely to continue to trouble the alliance as to the very nature of the threat and what to do about it are:

1. The change in the balance of power, which has heightened Western Europe's sense of physical vulnerability, and intensified allied anxieties about the reliability and effectiveness of U.S. protection and leadership.

2. The development of a degree of détente in Europe, which is regarded by many Europeans as both an inevitable step toward peaceful life on the European continent and a crucial counterpart to collective defense. It has given the Europeans a tangible stake in maintaining the status quo and in decoupling their relations with the Russians from tensions and crises outside the NATO area. The United States, on the other hand, is inclined to view détente as indivisible and to deplore a differential East–West posture that protects allied fruits of conciliation at the expense of conceding the Soviets a free hand outside the NATO area.

3. The projection of Soviet influence as well as indigenous threats to Western access to vital resources in the Third World—through arms aid, Cuban and East German surrogates, and Soviet forces—has greatly increased. Ideally, this magnification of a common security threat

to the United States and its allies, should have galvanized them to take immediate concerted countermeasures. In reality it has accentuated differences of interest and perspectives. It has put the United States in the position of assuming the burden and risks of containment outside the NATO area on behalf of allies that fear such action may jeopardize détente. This has led some of the allies to look for ways to insulate themselves from regional conflicts.

4. The submerged but profound differences over the need to improve the conventional posture of the forces of the alliance could lead to a major element of misunderstanding and distrust.

5. Burden sharing, particularly in light of the new tasks which the United States is undertaking in the Third World, especially for the Middle East, will be considerably in the public eye in days to come. As the American defense burdens increase and as Europe's GNP per capita continues to approach that of the U.S., there will be strong pressures for greater contributions from the European allies.

And there are other problem areas as well. For example, European opinion attaches greater importance to the continuation of the arms limitation process and appears far more optimistic than does American opinion. It is a political imperative in Europe to be able to demonstrate that all reasonable efforts along that line are being exhausted in order to justify heavy defense expenditures; while American opinion is far less insistent on that point. And economic and financial relations among the industrialized nations are central to all other aspects of cooperation.

Overarching these differences is the general tendency of European opinion to leave the military and political responsibility for security outside the NATO area to the United States, while at the same time remaining free to criticize Washington for undue emphasis on the military nature of the threat and on the military side of solutions. This clash can be alleviated to the extent that the European allies become more involved and more responsible in these extra-NATO matters and the United States increasingly shares responsibility and direction of these enterprises with its friends. However, to the extent that common cause is not made on these matters, then it will be doubly important that the Europeans react more constructively to America's exercise of the responsibilities Europeans do not wish to share. By the same token Americans will need to demonstrate a greater readiness to couple nonmilitary solutions with those of a military nature.

If these problem areas are not openly recognized, and persistent efforts made either to eliminate or at least limit them on the one hand, or to carry on as an alliance despite them on the other, then they could breed misunderstanding and distrust. It is this potential for increasing disarray that is so disturbing.

The allies cannot afford to overlook either the dangers of exacerbated differences or the risks of inadequate defense measures, whether military or nonmilitary. They cannot afford to give absolute priority to the need for agreement in their ranks, else they would be certain to attain only least common denominator agreements. Not one can afford to press for defense measures which would risk continued support for NATO. They must nevertheless find the means to improve their defense posture and share that burden equitably.

While it would be self-delusion to pretend that all differences in point of view can be resolved easily or speedily, experience indicates that patient consultation among allies can, over time, find common ground. Thorough consultation and collaboration have borne fruit for the allies in the mutual balanced force reduction talks (MBFR), where the allies have displayed remarkable unity on issues involving both defense policy and arms limitations policy. Similar unity among the Western nations at the Madrid conference monitoring the Helsinki agreements on the Soviet threats to Poland was the result of common appreciation of the importance of the issue.

If the allies could develop a procedure that would result in consultation being initiated even as early as the time when divergences were merely anticipated, early consultation could greatly reduce the possibility of the allies working at cross purposes. Reaching full agreement within the alliance on all such issues would be an unrealistic goal. There are issues on which no complete consensus exists or is likely to emerge. On the other hand, identifying the political risks of disagreement, when disagreement is inevitable, and working to minimize those risks, is well within the realm of the possible. The allies should agree to bridge the gap between differing views wherever possible, make every effort to avoid surprises for allies whose views may differ, and try to limit the scope and impact of remaining disagreements.

The question of allied cooperation must be seen and understood not only by governments, but by parliaments, by the media and by informed public opinion as part of a long term, ongoing process. In this respect the role of the North Atlantic Assembly (and particularly of the U.S. participation therein) should be strengthened. After all, we are talking of a reinvigorated defense effort reaching well through this decade; and the opportunities for disagreements over any such time frame are naturally high. The problem is not in disagreement; it is in whether or not diversity of view in an alliance can be channeled to work for the common good.

The genius of the civilization of the Atlantic nations has been the ability to strike a reasonable balance between individual liberty and responsibility to others. That concept, if applied to allied as well as domestic affairs, can afford the answer.

CONCLUSIONS

Six important messages result from the foregoing review of the significance for the Atlantic allies of the current manifold threats to their security.

First, the security of the allies can be endangered by events outside the NATO area just as much as by the threat in Europe, and by political warfare, whether at home or abroad, just as much by the military threat. The defensive measures of the allies, whether within or without the alliance, must be equally ecumenical.

Soviet military strength cannot be permitted to dominate any friendly region, whether in Europe, Asia, Latin America, or the Middle East. The armed forces of the free world should be able to deter and check Soviet attempts to subjugate free peoples by force or fear, whether the threat be direct or indirect.

The allies should equally be able to help cope with the use of militant surrogates of the Soviets to subvert or overthrow governments friendly to the West without getting bogged down in another Vietnam.

They must increasingly be prepared to successfully employ means short of armed force, such as political and economic countermeasures, to respond to threats short of actual hostilities.

In the Third World, they must work through and strengthen the independent governments politically and economically so as to help them resist external and internal dangers and seek to increase common interests with them.

Though the lead role may fall to the United States in defending the common interests of allies abroad, the resistance to Soviet expansionism must increasingly reflect a broad base of international participation and support.

Second, the United States must, by its actions and attitudes, reassume the global responsibilities of leadership among the free world nations that cannot otherwise be fulfilled. The response of the European allies thereto is equally indispensable for the continued confidence and strength of the alliance.

The retrenchment of American power and leadership in recent years has strongly affected allied confidence in and reliance on the United States. Reassurance will depend on insuring continuing general strategic nuclear parity and restoring a worldwide American military posture otherwise consonant with the requirements. It will equally depend on American ability and willingness to relate more closely to European thought and European ability and willingness to reciprocate.

The resolve of the United States will be judged by friend and foe alike by the nature and scope of its defense program, and above all, by the example in terms of willingness to sacrifice it will set for the alliance.

There is no reason why the European allies, particularly those which are now as well off as the United States, should not hold themselves to the same high standard of increased defense effort.

The public support necessary for a common defense effort still depends on alliancewide devotion to the concepts of self-help and mutual aid. If that exists all other differences can be dealt with satisfyingly. Without it the alliance will falter.

Third, deficiencies in conventional strength in a time when the West no longer has nuclear superiority must not be allowed to dangerously affect the credibility of the NATO deterrent strategy.

Unless there is a significant improvement in the readiness, interoperability, and sustainability of NATO's conventional forces, there is increasing risk that Soviet planners could come to doubt the purpose as well as the ability of the alliance to defend itself, whether at the conventional or nuclear level.

The improvement of the posture of NATO's general purpose forces is thus more urgently essential to bring the NATO deterrent up to date than ever before. In the absence thereof, loss of mutual trust and confidence within the alliance, as well as greater risks in East–West relations, will all too likely follow.

Fourth, the reservoir of military reservists in Europe should be tapped, first to provide a reserve for Allied Forces Central Europe and thus compensate for United States forces presently allocated to NATO which may be required to protect the interests of allied nations outside the NATO area, and subsequently to further strengthen the conventional capability of the alliance.

Affording greater assurance of timely arrival at reserve battle stations on the European front and at markedly lower cost to the alliance, common sense indicates that initial planning should cover the equivalent of the two to three divisions of American forces which might be required elsewhere.

In the longer term, additional combat units drawn from this pool could further enhance the conventional capability of the alliance on a most economical basis.

Fifth, the allies must be increasingly conscious of the risks of disagreements among themselves and together find ways to keep such differences to manageable proportions, recognizing that without political harmony there can be no common defense.

The augmentation of consultation on individual and collective measures to respond to challenges to their security should reduce the dangers of surprise among friends and the depth of disagreements among them. Nevertheless, divergent views of the European allies on the one hand and the United States on the other will continue to arise. The allies must increasingly be alert to anticipate such problems and to develop means to minimize their impact. Not one of them can afford the luxury of going it alone.

Sixth, allied cohesion, and the political will to contribute separately and jointly to the common defense depends upon public understanding of the issues involved.

The enduring strength of the Atlantic alliance lies, in the words of the Preamble of the Treaty, in the determination of its members "to safeguard the freedom, common heritage and civilizations of their peoples."

In recent years that determination has shown serious signs of erosion, particularly on the part of the younger generations which have no personal memories of the causes, suffering, common effort, and results of World War II.

POLICY PAPER. THE CREDIBILITY OF THE NATO DETERRENT   45

Keeping that determination strong from generation to generation calls for stimulating the appreciation throughout the peoples of the alliance of the basic values of Western civilization and how they can be preserved.

RECOMMENDATIONS

The nations of the Atlantic alliance must jointly expand their concept of defense and their activities, whether within or outside of NATO:

1. They should improve their procedures and their practices for identifying challenges from any quarter to the security of any of them, and for consulting effectively on measures to be taken individually or in concert in response to such challenges.
2. Such procedures and practices should increasingly embrace the development of political and economic as well as military measures in response to challenges short of the use of armed force.
3. They should develop the means to work more closely with the nations outside of NATO whose security is threatened by external forces, and should consult more closely with them and with each other on such challenges.
4. They should give priority to the task of identifying and controlling the dangers of divergent perceptions and approaches within their ranks equal to that of dealing with external threats.
5. They should increase efforts to develop greater awareness, particularly among the young, and among those in the armed services, of the common need to strengthen and defend the basic values of Western civilization.

They should equally improve their means of defense, and jointly consider the following proposals:

1. Expand military capability so that challenges outside the NATO area can be met without affecting the credibility of the NATO deterrent.
2. Compensate for the contingent need to use perhaps two to three American divisions now earmarked for NATO in other

areas by drawing on the reservoir of readily available European reservists and civilian resources to form equivalent replacement units.

3. Provide the means to establish the readiness of NATO's conventional forces by providing the equipment, manpower, and training now in seriously short supply, as specifically noted in this study.

4. Increasingly accord to the role of conventional forces the priority required of them as the result of the passing of the period of Western nuclear superiority.

5. Proceed to deploy modern long-range theater nuclear weapons while continuing to seek satisfactory arms limitation agreements.

6. Be willing to join our allies in pressing for arms control agreements which will enhance stability, particularly confidence-building measures.

7. In the longer term, increase the conventional force level by organizing trained European manpower into additional reserve units.

8. Continue support for and the ability to reinforce the countries of the northern and southern flanks: particularly the political, economic and military assistance needed by Turkey to play its full role in the alliance.

9. Seek to continually adjust the burdens of global defense so that nations which benefit from efforts in the common cause share more equally in the burdens thereof.

10. Make a major effort to break through the resistance to developing a more economic use of the defense production resources of the alliance as a whole.

Specific recommendations for the United States, some of which could well apply to other allies, are:

1. The United States must demonstrate its readiness to resume a role of responsible and consistent leadership in world affairs.

2. It should increasingly demonstrate its readiness to appreciate and consider the particular concerns and viewpoints of its allies,

expecting only that the allies equally appreciate and consider the concerns and viewpoint of the United States.

3. The United States should stimulate a new conviction among allied and friendly nations that, with full collaboration, it will be possible both to preserve liberty and keep the peace.

4. American leadership by example is needed to stimulate allied confidence and accomplishment in restoring the balance of power.

5. The Congress and the executive branch must collaborate to provide the basis for a long-term and consistent nonpartisan foreign policy and defense program for the years ahead.

6. The United States should recognize that progress toward self-sufficiency in energy, toward financial stability and economic growth are indispensable if the United States is to meet its international obligations and its defense needs.

7. Not only must the central strategic forces be strengthened, and the long-range theater nuclear force be supported, but American general purpose forces must be provided with the equipment, the manpower, the operating and maintenance budgets, the war reserve material, the logistic and air and sea lift support and protection, and the industrial mobilization base that will materially increase their ability to sustain a conventional defense. These revised goals should be clearly set out and the long-range programs to realize them projected, to insure the steady progress toward these ends.

8. The United States should take the lead in seeking to improve the rationalization, standardization, and above all, interoperability of defense equipment of the allies, seeking to develop in practice a reciprocally beneficial system that would enhance collective allied strength and at the same time abandoning the more nationalistic approach it has in fact too long pursued.

9. The United States should give high priority to improving its military manpower posture, and be prepared to turn to some form of compulsory service if adequate results are not forthcoming from other means in the near future.

10. The United States should provide increased and more consistent support to bilateral and multilateral economic assistance pro-

grams for the developing countries and maintain its leadership in seeking to open the markets of the world on a reciprocal basis.

11. The United States should support a wider effort in the field of foreign information broadcasts and publications so that the people and the governments of the developing countries are more fully informed about world matters.

SPECIAL COMMENT — *Leonard Sullivan, Jr.*

I agree with the central themes of this report, but believe it seriously underestimates the severity of NATO's inadequacies. It ignores the clear and evident need for major restructuring of the strategies, roles, and missions of both NATO and the larger Western alliance.

The Working Group has begged the basic issue of whether or not, within the context of the unfavorable worldwide shifts in the "correlation of forces," NATO presents a credible deterrent to Soviet bloc exploitation. Like NATO itself, this group fails to differentiate between what might deter the West and what will deter the Soviets. I find it highly unlikely that the Soviets are *militarily deterred* by a European NATO that:

- Has no mechanism for the tactical or strategic linkage of military security threats in different parts of the world;
- Depends critically on costly U.S. transoceanic reinforcements, also earmarked for other contingencies, rather than developing its own more effective latent mobilization potential;
- Counts on nuclear escalation to avoid running out of conventional munitions;
- Is composed of disparate multinational forces which do not use common equipment, cannot operate together, and could not resupply each other;
- Fails to prepare the battlefield, and does not intend to defend itself in depth;
- Will not inconvenience itself sufficiently to share equally in the burdens of its own defense; and
- Assumes that the United States will absorb the predominant burden of nuclear destruction to avoid damaging Europe.

The Soviets are, at virtually no risk, expanding their political and economic inroads into Western Europe and encouraging the degeneration of the will and solidarity of the NATO members. How are they deterred from the pursuit of their objectives? Why on earth should they exercise military force unless or until Europe resists their advances?

If, on the other hand, Western Europe does eventually find the will to resist the bear's hug, then they will certainly first have to eliminate the strategic military absurdities noted above. There is no shortage of innovative alternatives available.

Equally as important as these military realities, the West must recognize that its ultimate long-term political and economic strength lies in the depth and diversity of its civil sector, not its military structure. Unless and until the West comes to grips with the fundamental issues of preserving its *collective* economic strength and learning to *use* it as an instrument of its overall security objectives, the chances of arresting the deleterious shifts in the total correlation of forces are slim indeed.

Finally, this paper tacitly implies that the "vital security interests" are essentially independent of the level of allied acceptance of responsibility. But what delusion should *any* member of NATO or the entire Western alliance be entitled to exploit the full benefits of our political and economic interdependence while inadequately contributing to our collective security? The large and real inequities in security burden-sharing are already a disgrace and appear destined to become worse. It is past time for the United States to recognize and assert that it cannot and will no longer do more than its share. If greater participation is not forthcoming from either our North American, our Atlantic, or our Pacific allies, then the United States must for its own good inilaterally readjust its own security objectives and strategies to bring its expectations into line with its capabilities over the long haul. We have no supernatural mandate to sacrifice our own future on the altar of allied indifference.

2 THE USSR AND THE WESTERN ALLIANCE

William G. Hyland

The USSR has pursued reasonably clear objectives in Europe: It has aspired to, achieved, and maintained a dominant position in Eastern Europe that it considers vital to Soviet security. A major political landmark in the Soviet Union's continuing effort to maintain its position there by whatever means necessary was reached in August 1970 with the signing of the Federal Republic of Germany treaty (FRG), whereby West Germany in effect accepted the political division of Germany into East and West. The advanced Soviet position in Eastern Europe has also been the foundation for pursuing the second Soviet objective: gaining significant influence over the policies of its adversaries, the principal Western European members of NATO and the United States. Since the early 1970s the USSR has used a relaxation of tensions to advance this aim. In doing so the Soviets have also sought to promote a third aim: to restrain and to weaken the military position of NATO through both negotiations and a substantial buildup of their own forces, particularly on the central front. And since the early 1970s the Soviets have begun to explore more actively a final objective: the possibility of dividing the Europeans from the Americans by creating a selective détente with Europe, designed to be immune from the fluctuations of Soviet–American relations. The chances for Soviet progress in Europe, however, have once again been threatened by the recurrence of challenges to the Soviet position in

Poland and, by implication, throughout Eastern Europe. These threats plus an extraordinary international turbulence at a time of a Soviet leadership succession all conspire to give Soviet policy a singularly fluid character.

EASTERN EUROPE

Moscow's determination to create and consolidate an empire in East Europe was an obvious consequence of the Soviet military victory in World War II. A number of Western observers during that war foresaw that the USSR would want to emerge as the "only important military and political force on the continent of Europe. The rest of Europe would be reduced to military and political impotence."[1]

The immediate Soviet aim was to use its dominant position in Eastern Europe as the guarantee against a German revival. Indeed, the Soviet fear of Germany runs throughout the wartime and postwar negotiations; and there is an air of fatalism if not resignation in frequent Soviet predictions of a German resurgence within 20 years or so. In order to achieve their minimal ambition of dominating the area, along a line coinciding roughly with the furthest point of the advance of the Soviet army, the USSR was willing to pay a political price. The price was, first of all, Western goodwill at the end of the war and, more important, limits on their ability to use Eastern European positions to influence Western decisions. Thus the irony was that in securing their position in Eastern Europe (including the abortive attempt to secure West Berlin), the Soviets inevitably produced the result they feared the most: the revival of Western Europe, the rehabilitation of Germany, and the return of American power to the continent. It so happened that a critical phase in the Western resurgence came in the period between the death of Stalin in March 1953 and the Geneva summit in 1955. This was probably the last real chance for a negotiated settlement based on a united Germany; it passed, and both sides began the laborious and painful process of accepting the division of Europe. It only remained for the uprising in Hungary and its suppression by the Red Army to make this tacit acceptance an explicit fact.

It may be worth quoting a reflection of the fundamental Soviet attitude (from a relatively new source). Before the Hungarian uprising, in May 1956 Khrushchev explained his attitude to the Yugoslav ambassador:

> It is interesting that, when he talked about the "camp" though in fact only about Poland, Khrushchev spoke of it as though it were an internal Soviet affair and not as though it concerned another country and the Soviet Union's relations with another state. Khrushchev said that they would not permit any innovations in this connection and they would oppose anybody who tried to introduce changes into relations between the Soviet Union and the "camp."[2]

To sum up, the first and foremost Soviet objective in Eastern Europe was, and remains the creation and protection of a security *glacis* in East Europe.

WESTERN EUROPE

The period that resulted in a German and European settlement 1970–1975 began with the shifting currents of political opinion in Germany in the middle 1960s. The change in the German attitude toward Moscow brought on by reaction to the Berlin wall was reinforced in part by U.S. beliefs that in the post–Khrushchev period it might be possible to develop an "alternative to partition" (to use Zbigniew Brzezinski's phrase). For centuries, Chancellor Kurt Kiesinger said in December 1966, Germany was the bridge between Eastern and Western Europe and now in our time we would like to fulfill this mission.

Within Germany the policy of "little steps" was adopted, in part to circumvent direct negotiations with the German Democratic Republic (GDR), to play on the loosening of Soviet–Eastern European ties, to exploit the opportunities for economic penetration, and, finally, to align West German policy with a broader European interest in détente.

In addition to the groundwork laid by West German policy during the period of the grand coalition and later under the decisive initiatives of the Social Democrat party government of Chancellor Willy Brandt, three other factors contributed to the evolution of a European settlement. The first was growing Soviet apprehension over China, an apprehension greatly intensified by the border clashes of 1969 and the Sino–U.S. rapprochement of 1971. It is likely that the opening of China encouraged the USSR in the Berlin negotiations during the summer of 1971, which in any case were a key to any further progress in Europe. Second was the impact on Moscow of the threat of further post-Czechoslovak troubles in Eastern Europe as manifested in the Polish riots of December 1970 and the consequent removal of Wladyslaw Gomulka, first secretary of the Polish Communist party. The internal Polish tensions that erupted in the immediate wake of and despite Gomulka's successful negotiation of a treaty with West Germany were sharp reminders that "selective détente" in Europe could not substitute for a broader based relaxation of tension that would have to include the United States. Thus the third factor was détente with the United States, which was formally inaugurated at the May 1972 summit and which had been made possible by the Berlin settlement of September 1971. The Berlin agreement had also opened the door to the broader European conference at Helsinki that was intended to crown party chairman Leonid Brezhnev's "peace program," announced at the 25th Communist Party Congress in March 1971.

The Helsinki conference, which convened only after tortured maneuvering on both sides, remains a source of controversy, especially in American political mythology. On the one hand Brezhnev achieved what had been denied to Stalin, Molotov, and Khrushchev: a pan-European acceptance of the territorial status quo in Europe. This was possible because territorial revision had long since been given up in the West. On the other hand, at Helsinki it was not Soviet influence that was predominant but American and Western influence. The psychological balance was evident then and in the subsequent defensive Soviet performance at the follow-up meetings.[3]

Moreover, the settlement inaugurated at Helsinki did not prove to be the means for developing more durable Soviet institutional or other ties to Eastern Europe. Indeed, Romania has openly treated Helsinki and subsequent meetings as the means for fending off the USSR; and other Eastern Europeans have taken refuge in the meetings to carve out a small degree of autonomy. The opposite effect has also been true, however. American and European attitudes toward Eastern Europe have mellowed, as reflected in the improvement of bilateral relations between the United States and Hungary and even Czechoslovakia.

It is impossible and perhaps erroneous to trace a direct line from Helsinki to the Polish crisis in the 1980s. Some observers believe that the Polish free trade unions could not have survived for long had it not been for a period of détente in Europe.⁴ The Madrid Conference for Security and Cooperation in Europe (CSCE) may have helped make the USSR reluctant to intervene. In any case there are profound consequences for Soviet power implicit in the appearance of free institutions in a Marxist–Leninist state. Clearly, such heresy is unacceptable over any long term. Regardless of the tactical expedients adopted in the crisis, the USSR has to emasculate or suffocate the political power of the new solidarity forces or to rout them with military force. In any case the message seems clear: Détente is not a sufficient condition for the flourishing of genuine ties between the Soviet Union and Eastern Europe. External atmosphere and even international treaties cannot substitute for the harsh realities that the internal systems of the states of Eastern Europe are essential failures. The fragility of the "Eastern bloc" thus remains a sharp limit on the diplomacy of the Soviet Union especially in regard to developing policies for European security and elaborating European détente.

DIVISIBLE DÉTENTE

While the Soviets have had "defensive" aims in using a relaxation of tensions to try to secure their East European dominions, this climate has also had "offensive" advantages. First and foremost the atmosphere was conducive to drawing Western Europe toward the Soviet sphere in ways that would provide a hedge against a subsequent deterioration in relations either outside of Europe or between the superpowers. One means of pursuing this aim has been increased economic links. Trade, credits, and a web of economic interrelationships between East and West had long been thought by many observers to be a major Western vehicle to influence the course of internal Soviet–Eastern European developments and to help shape Soviet foreign policy as well. In the 1960s in particular it was a strongly held view that Soviet interest in economic links could be turned to Western advantage. Indeed, trade was considered by some to be a wedge for promoting a relaxation of tensions. This view eventually clashed with another school that favored "linkage," joining trade concessions to political concessions from Moscow. The ultimate linkage was to join trade with internal Soviet political changes as reflected in American demands for free emigration from the USSR. But this American debate had no resonance in Europe. In fact, Soviet overtures for a new economic relationship found fertile soil and a policy gap was thereby opened between America and its NATO allies.

In the five year plan of 1971–1976, the Soviets began to build into their economic expectations a higher level of trade with the West, including imports of high technology to be financed largely by Western credit. Soviet desires in this regard were never concealed, and in fact one of the leading edges in the Soviet peace offensive of the 1970s was the economic offer.

The Soviets succeeded handsomely. In 1973, for example, Communist trade with European NATO countries was about $19 billion, but within three years it had grown to over $32 billion; it was over $50 billion by 1979. Exports to bloc countries grew from $10 billion to over $18 billion, almost doubling within 3 years. The USSR was the dominant partner, of course ($26 billion in 1979). Interestingly, Soviet trade with West Germany overtook internal German trade in that FRG exports to the Soviet Union surpassed FRG exports to East Germany in 1973. Soviet exports came to include considerable energy resources—natural gas and some oil.

The expansion of trade had to be financed from Western resources, since the USSR has historically had a chronic shortage of hard currency and has adopted a policy of earning hard currencies by limited sales of gold. In effect, the USSR was able to borrow heavily. In 1971 the Soviet hard currency debt to the West was about $1.8 billion, not including shares in CEMA bank borrowing. By 1979 it had grown to over $17 billion, the greatest leap taking place in 1974–75, that is, coinciding with the energy crisis and Helsinki conference. The statistics for the Eastern bloc as a whole are even more dramatic: growing from $8.5 billion of hard currency indebtedness in 1971 to over $77 billion in 1979. It is also important to note that this debt is not all from governmental sources but includes a large share from commercial lending in the private sectors, thus drawing in a particularly important segment of the political elite in Western countries.

Trade and credit were the material underpinnings for the political relationship that came to be known as "divisible détente." The Soviet tactic was to treat the principal Western European countries separately from the United States. The Soviets correctly calculated that the internal political dynamics of Western Europe, especially the existence of Communist parties and strong left wing factions of non-Communist coalitions, meant that the governing parties would have to be more sensitive to the domestic implications of strains in the international atmosphere. Moreover, the Soviets also calculated that Europe in the wake of abortive negotiations with the United States for the "Year of Europe" would be open to opportunities to assert its independence from the United States. The Middle East, for example, was an area where European and Soviet interests tended to coincide—in supporting the Palestine Liberation Organization (PLO) and urging a settlement through the large negotiations at Geneva. The overall Soviet aim in the 1970s was to create a political impression in most of Europe that détente was a permanent feature that could be relied on, and it remains so; this is as Chancellor Helmut Schmidt put it, "a policy of calculability on both sides." Brezhnev himself described the Soviet view in late 1977:

> The salutary changes in the world which have become especially appreciable in the 1970s have been called international détente. These changes are tangible and concrete. They consist of recognizing and enacting in international documents a kind of code of rules for honest and fair relations between countries.... They consist of a ramified network of agreements covering many areas of peaceful co-operation between states with different social systems. The changes for the better are most conspicuous in Europe.[5]

Brezhnev's appreciation was to some extent shared in Western Europe. The importance of trade, for example, was recognized by Chancellor Schmidt in a lecture before the Institute for Strategic Studies in October 1977, when he described the world recession of 1975 in which German exports fell by 4 percent but exports to the Soviet Union rose by 46 percent: "thus making a valuable contribution towards improved use of capacities and a better employment situation in my country." Trade with the East, which had been conceived and justified as a weapon for opening up the Soviet Union and shaping its policies, had clearly become a double-edged sword. Indeed, as Schmidt pointed out, it was questionable which side benefited the most, but if Western countries acted jointly, the development of trade with the East could "be essential for both our own economic security and the safeguarding of peace."

The impact of détente was also psychological and evolved to the point that the Western Europeans could say before the invasion of Afghanistan (again quoting Helmut Schmidt) that "we are living in more secure circumstances than in the first 25 years after the war."[6] Thus, matters gradually reached the stage that whereas the American commitment to détente was tentative and subject to reversal, particularly after the Angolan conflict of 1975, the European commitment continued to grow. One observer wrote:

> The United States and Western Europe have evolved different kinds of relationships with the Soviet Union, and this is behind the trouble in the alliance. ... There are many (in the US) who oppose improvement. It is different in Western Europe, where every country, with the exception of Britain, has strong constituencies committed to détente ... Détente has become a discredited word in America. For the vulnerable Europeans, however, détente has meant organic improvement in their lives—politically, militarily, and economically.[7]

DÉTENTE WITHOUT SECURITY

The consolidation of the political status quo at Helsinki included no provision for a parallel military relationship in Europe. The connection between the Helsinki conference and the substantive negotiations on mutual force reductions had been grudgingly accepted by the Soviets, but in practice the MBFR negotiations had lagged well behind those at Helsinki. The slow pace reflected more than technical, tactical differences. The Soviets faced a major dilemma in Central Europe: Ending the German menace also meant giving up the cohesion inherent in the invocation of a constant external threat to Eastern Europe; obviously one could not preach German revanchism while signing treaties with the West German chancellor in Moscow. On the other hand, the Soviets would undermine their own policies if, at a time of shoring up the territorial-political status quo, they began to undo it through the destabilizing process of withdrawing Soviet troops from Germany or the central region of the Warsaw Pact. But, finally, the Soviets could not expect to continue pursuing a European détente while their military plans proceeded on an independent plane, particularly since they had decided to modernize their medium-range strategic forces in European Russia.

Soviet policy turned out to be a blend of propaganda, abstract theory, and tactical dexterity. First, the Soviets adopted immediately the slogan of supplementing détente with a "military détente." Second, they developed the theory that the proper outcome of the

Vienna MBFR negotiations would be to reflect the existing balance, but at a lower level. And finally, they took refuge in a Western-inspired debate about the proper counting rules for including military forces in MBFR. All of this simply meant no progress.

The Soviets did provide some indication that they might be interested in a token or symbolic reduction. But in general they insisted on "equal security," that is, to treat the negotiations as an exercise in ratifying the existing relationship of forces in the area, without regard to other negotiations, such as the Strategic Arms Limitation Treaty (SALT), or to military developments in the USSR. In adopting delaying tactics, however, the Soviets eventually had to pay a price in terms of dwindling influence over Western plans. At first the Soviets scored gains simply because the fact of MBFR negotiations had a restraining effect in NATO. The Western participants were reluctant to confront the problem of modernization or reinforcement lest the Vienna negotiations be jeopardized. And as the stalemate deepened, the Western powers posed various sweeteners, offering to include the withdrawal of U.S. F-4 aircraft and Pershing missiles from Germany, on the theory that restraining U.S. nuclear deployments was appropriate compensation for larger Soviets reductions in conventional arms and manpower. As long as this proposition was in play at the negotiations, the Soviets were under no pressure to reach agreements or restrain their own plans for modernization and rebuilding their medium-range missile forces. Indeed, the general thrust of Soviet tactics was to respond slowly by offering just enough to prevent a showdown in the negotiations.

By 1977-78, however, it was more and more evident that the Vienna negotiations were being used as a political cover for a new Soviet buildup. The endless debates about which side had the correct data were irrelevant to the fact that the Soviets were deploying a new mobile intermediate-range missile with a multiple warhead as a replacement for the older SS-4 and SS-5. It became increasingly evident that the "Euro-strategic balance" was a victim of neglect in the Western negotiating positions (the SS-20 was excluded from both MBFR and SALT II); so that leading European statesmen began to express concern that the European balance was endangered and some adjustment was necessary either by new Western deployments or by new negotiations or a combination of both. This was the thrust of Helmut Schmidt's Buchan memorial lecture at the International Institute for Strategic Studies in October 1977.[8]

The subsequent campaign by the Soviet Union to head off a NATO decision to deploy cruise missiles and Pershing II in Europe illustrates both the limits and strengths of Soviet policy: first, the inability of the Soviets to develop the line of a genuine détente in Central Europe without offering some military concessions and, second, the inhibitions within Western Europe against confronting the Soviet Union with the consequences of its military buildup.

The USSR under increasing pressures from the West, made a series of overtures and offers in 1978-1980, to stop new Western missile deployments. A major trade agreement was arranged with Bonn; the Soviets accepted the concept of a common ceiling in MBFR (thereby touching off the hopeless quarrel over numbers); there was a successful visit by Andrei Gromyko to France in October 1978, and the usual laundry list of Soviet proposals at the United Nations and the Warsaw Pact meetings. A new Soviet MBFR proposal was designed to take up the Western offer on withdrawal of nuclear systems in return for a Soviet withdrawal of 30,000 troops and 1,000 tanks. This was intended to put the West in a bind: If the Western powers accepted their own position, the effective result would be a freeze on nuclear systems in Germany and the Benelux countries. Cruise missiles would be handled in SALT II, and the West would thereby be effectively deprived of a military response to the Soviet buildup.

It must be noted that at this point Soviet policy outside Europe was moving into a sharply more aggressive and activist phase in Zaire, Ethiopia, Aden, and in Afghanistan. In some part this was a reaction to the growing fears of a new encirclement by the United States, Europe, China, and Japan. This phenomenon plus the threat of a revival of Western theater nuclear armaments reinforced the traditional policy of the USSR in the face of growing threat: namely, to intensify their armaments and to intensify their public pressures. Brezhnev, for example, had upgraded the value of MBFR to "today's priority task.[9] And after the U.S.-Soviet summit of June 1979, Gromyko pressed for "some curtailment of U.S. and Soviet troops alone," which he claimed would be a "great achievement." But he also acknowledged that the talks were making "no noticeable progress."[10]

The impact of SALT and the Carter-Brezhnev summit, however, was to demonstrate that European interests continued to fall outside the negotiating framework. Cruise missiles were to be restrained in Europe until the end of 1981 and perhaps longer if the SALT protocol were to be extended. Soviet Backfire bombers and SS-20s, however, were effectively excluded from SALT. As a result the Europeans were increasingly determined to create a better military balance in Europe. Despite dramatic offers to limit Soviet medium-range missiles and even the announcement of unilateral withdrawal of 20,000 Soviet troops from Germany, NATO could not be deflected from its determination to redress the military balance, with or without negotiations. And this was the decision of NATO in December 1979. It was a sharp setback to the Soviet strategy of achieving détente without offering security.

But Soviet diplomacy and the harsh threats issued before the NATO meeting and after did have an effect. The major countries involved, Germany, the Benelux, and England, began to diverge. In order to placate the SPD left wing, Germany insisted that one other continental member of NATO agree to accept the American cruise missile deployments. But then the smaller countries, Belgium and the Netherlands, insisted that the deployments had to be accompanied by an attempt at negotiation. This meant that the entire NATO position would be extremely vulnerable to Soviet tactical exploitation. And indeed, despite the Afghan crisis, the Soviets gradually came to appreciate this vulnerability and abandoned the harsh threats of early 1980; through contacts with Chancellor Schmidt they opened the way to preliminary negotiations, which began on October 13, 1980, but ended without progress.

In any case the USSR has a tactical advantage. Even if negotiations are protracted but still offer some slight hope, it is unlikely that the American cruise missile deployments will proceed in either Belgium or Holland. Hence it may become politically impossible for the Germans to proceed. And in this situation the chance of developing a negotiated compromise shaded in favor of the USSR becomes increasingly likely. In any case, the Soviets appear determined to forestall the American buildup of military force in Europe and to play heavily on European fears of a new cold war in order to check NATO rearmament. They have two or three years before planned of deployment for the new missiles. At the Communist Party Congress, Brezhnev revived the campaign against the NATO deployments; he not only repeated the moratorium on SS-20 deployments but offered a new proposal to extend advance notification of maneuvers to all of European Russia.

THE INDIRECT THREAT: SOUTH ASIA AND THE MIDDLE EAST

As noted, in 1977-78 the Soviets began to see the specter of encirclement arising once again, this time anchored in Western ties to China and Japan. In the aftermath of Chairman Mao Tse Tung's death the Soviets had made an effort to open the door to a rapprochement with Mao's successors, but to little avail. By Spring 1978 it was clear that the Soviet effort was not only failing but that the other powers were coalescing: Japan and China, encouraged by the United States, signed a treaty in August following a successful visit to China by President Carter's national security adviser, Zbigniew Brzezinski. And the Chinese leaders embarked on a unique foreign tour of Europe and South Asia (ironically including Iran). There was extensive speculation about European arms sales to China and, of course, growing talk of Japanese rearmament.

This was the background against which the USSR in 1978 had intervened in Yemen and Afghanistan to assist Communist coups, had given the North Vietnamese a blank check in their treaty in November 1978, and had drawn the new Afghan regime closer to the Soviet orbit in a state treaty of December 1978. Then, of course, events rescued Soviet policy at least for a time: The fall of shah of Iran opened a gaping hole in the northern tier. Afghanistan was falling under an outright Soviet-style regime, while Iran was moving against American interests.

Whether by design or not the Soviets began to assert a predominant interest in South Asia to the exclusion of the United States. Thus, in November 1978, Brezhnev laid claim to an equal security voice in Iran, an area that had been allied to the United States for over 25 years. The deterioration with the United States was more and more frequently explained in terms of a new balance of power. Thus journalist James Reston was told by officials in Moscow in November 1979 that "The core of the problem ... is that the United States seems unable to realize that it can no longer have its own way in the world as it did in the years just after the Second World War, but is now vulnerable like other nations and compelled to come to terms with conflicting political, economic and military forces."

The capture of the American hostages in Iran followed so quickly by the Soviet invasion of Afghanistan touched off a major crisis. In

Europe a war scare seemed imminent; even the redoubtable German chancellor saw signs of 1914.

A central feature of the 1980 crisis over Iran and Afghanistan was the growing divergence in NATO over fundamental elements of Western policy, that is, the failure to agree on an assessment of Soviet policy. Increasingly in Europe, Afghanistan was written off as an accident on the road to détente or a preventive measure, while in America it was seen as a watershed opening a new dangerous era of confrontation. The willingness to respond was hence different in Europe and America. The Europeans offered formulas for neutralizing Afghanistan, certainly without any genuine hope of success, whereas the United States concentrated on punitive measures, also without much hope of forcing a Soviet withdrawal. The French asserted their right to an independent policy on matters of European détente, and Germany acted independently without asserting the right to do so. Divisible détente had reached a dangerous point. The tension was broken in part by Chancellor Schmidt's visit to Moscow and his return with an unconditional offer to negotiate on theater nuclear arms (treated as a concession in Germany). Of course it was typical effrontery for the USSR while engaging in a major buildup of SS-20s to make a Western moratorium a precondition to negotiations. And the difficulties the Soviet occupation army encountered in Afghanistan eased the tensions in the West as did the long period of American inaction on the issue of the American hostages held in Iran after the failure of the rescue mission in April 1980.

Nevertheless the net effect of Soviet moves in 1980 was ominous for the Atlantic alliance. The intervention in Afghanistan was a deliberately calculated decision to use Soviet troops, by a leadership that had usually been thought prudent and not inclined to risk. In Afghanistan the Soviets had acted, perhaps because they had run out of viable political options, but also because they no longer considered direct intervention in Afghanistan a risky, dangerous policy. Indeed, what Moscow saw was a new regional instability on its southern frontier, set against a new global balance that gave the Soviet Union a new freedom of action. Not only did they see new opportunities opening in the Third World as highly tempting, but they saw Western reactions growing less and less effective, including American resolve to counter Soviet advances. And finally the Soviets drew the conclusion that the "relations of international forces and especially of international military power, have changed drastically in their favor."[11]

The consequences of Afghanistan and Iran are still being played out. The Iraqi attack on Iran reflected not only an age-old settling of accounts but a further breakdown in the international system and another indication of the growing irrelevance of American power to the revolutionary turmoil of the Third World. The tentative emergence of the Soviet Union as the dominant power in South Asia and the Persian Gulf unleashed a diplomatic scurrying for new alignments and positions and the bizarre situation of American allies, Jordan and Saudi Arabia, supporting a Soviet client, Iraq, while Libya supported Iran, and Syria sought reassurance from Moscow. All of this was occurring in an area of vital importance to European nations, who consequently were being forced to consider a more dynamic role in global affairs if their interests were to be protected against chaos and the advance of Soviet power.

THE AMERICAN CONNECTION

The Afghan and Iranian crises revealed two important areas of conflict within the Western alliance: (1) how to meet the Soviet problem, whether by continuation of a limited détente or by deliberate interruption of relations and revival of counterpressures and (2) how to manage the regional turmoil, whether by placating the radical and revolutionary elements as the Europeans were inclined or by drawing a line and reinforcing the moderates, a position the Americans were groping for in the Carter doctrine. These divisions gave the USSR scope for action in playing to the European options. Although explainable on tactical grounds, Soviet action nevertheless raised an old question: Does the USSR want the United States out of Europe?

For much of the postwar period, it was obvious that the Soviets wished to see the Americans withdraw from Western Europe. Though encouraging various disengagement schemes in the 1950s and early 1960s, the Soviets could not offer the one thing that might have precluded the American presence: a total Soviet withdrawal. Gradually, with the revival of West German power the Soviets began to reexamine the proposition of forcing the Americans out. Especially after the Czech crisis died down the Soviets increasingly had to contemplate the risks of increased pressure for a mutual withdrawal, as had been proposed by NATO on the eve of the Soviet invasion of Czechoslovakia. Indeed, the pressure of Vietnam was leading to a unilateral American drawdown. The net result from the Soviet perspective was that Germany loomed larger and larger in Western policy; moreover, the internal German disputes had been papered over in a grand coalition bent on pursuing a national policy.

American forces therefore were still a balance against Bonn and a legitimation of the retention of Soviet forces in West Europe and were not being significantly strengthened. Matters even reached the point that in the midst of a major American debate over the Mansfield amendment to withdraw 150,000 American troops from Europe, Brezhnev intervened with a gratuitous offer of negotiated troop reductions. This surprise helped to turn the debate around. Its timing is still something of a mystery; it may be explained by the Soviet desire to ensure against the China opening and to build on the political settlement reached with Brandt by preserving the military status quo.[12]

The suggestion that the USSR actually preferred an American presence was also reflected by the agreement that the United States and Canada could participate in the European Security Conference. And this general policy persisted, until Fall 1979 when it may have come under review. Of course, it is not evident that the Soviets now want to force an American withdrawal. There is no specific proposal. But Soviet conduct raises some interesting speculation.[13] First, the Soviet Union ostentatiously withdrew some troops from Germany, beginning with Brezhnev's announcement in October 1979, which had been preceded by a Soviet suggestion of a small Soviet–American withdrawal. Second, the Soviets insisted that the American nuclear deployment be halted and then negotiated. Certainly the Soviet aim is either to eliminate the new American nuclear deployments or reduce these to the level of symbolism.

In other words the USSR confronts Europe with the fact that the conventional imbalances will not be reduced nor any real limits placed on Soviet weapons, except as part of American disengagement. And this raises the question of whether NATO's strategy could be maintained when the strategic balance is no better than parity. In such strategic circumstances the reduction of American power on the continent would be greatly magnified. Even if the SS-20 were to be limited the maintenance of the current NATO strategy requires a substantial American deployment in theater nuclear weapons. Thus negotiations will eventually require NATO to make a decision of profound long-term significance: whether to seek reassurance and security in limited arms control arrangements or to place greater reliance on maintaining a balance of military power. The choice in the 1980s is by no means clear on either side of the Atlantic.

FINLANDIZATION

One reason it may be that Soviet policy intends to drive a new wedge between America and its Western European allies is the prospect that through a combination of factors, the Soviets could gain decisive influence over key Western countries. In the past this was regarded by even the more pessimistic observers as a pipe dream of the Kremlin. The sophisticated Europeans would not be entangled in such a snare and delusion as reliance on the goodwill of Moscow. Moreover the Europeans had economic leverage and a not insignificant military position to counterpose against Soviet pressures. Finally there was and is the American guarantee.

Each of these major factors has become more debatable. First, the European's sophistication about the Soviet Union is tempered by a growing sentiment that both of the superpowers are almost equally at fault in managing the global crisis. The European left wing in particular is much less impressed by the Soviet threat than by the increasing accumulation of Soviet problems. The insistence on the Afghan quagmire is the most recent rationalization of the Soviet threat.

Second, as already noted, the economic relationship is no longer so clearly one-sided. The fact is that both Germany and France are becoming more and more dependent on Soviet supplies of energy, as well as a high volume of trade. The issue is not whether these countries could withstand blackmail but whether they could withstand the more ambiguous graymail that is more characteristic of Finlandization. To some extent such a contingency depends on events beyond European control: namely, in the Middle East and Persian Gulf. And that is why the threat of advancing Soviet power is as critical to Europe as to the United States. The Soviets, in effect, are developing two prongs: threats of manipulated trade plus a cutoff of oil by pressure in the Persian Gulf. The ability of Europe to weather a creeping crisis of this sort is at least open to debate. And it is of more than passing interest that in this connection the Soviets have hinted that a joint Soviet-European policy could guarantee the vital oil supplies. This, of course, is a definition of Finlandization.

The growing military imbalances must be treated separately. But the political impact is a critical element in the future of the Atlantic alliance. The pressures of extra-European crises on American forces and resources are growing with each new crisis. The strong implication is that the Europeans will have to share more of the security burden, in some undefined ways, but almost certainly by increasing readiness, stocks, and perhaps, force levels in Central Europe. Some U.S. naval and air requirements currently met by U.S. forces will have to be shared, say in the Mediterranean, the Indian Ocean, or the North Atlantic. In any case, it is evident that the United States would not be able to engage in warfare in Europe and outside it without assistance and that the Europeans cannot risk leaving decisions on extra-European commitments to the United States alone. What, then, is the consequence for Europe of assuming more of its own direct defense in the face of a new balance of power in which Soviet domination looms more and more ominous? At a minimum it increases Europe's vulnerability to Soviet pressures, not only the traditional pressures in Central Europe but pressures in other areas important to Europe's economic survival.

THE OUTLOOK

A consensus seems to exist that the next several years will be critical for NATO's relations with the Soviet Union. Soviet strategy in the early to mid 1980s will be determined by certain strategic facts of life. The most important is that all of the major power centers— Europe, Japan, China, and the United States—are to some degree hostile to the Soviet Union, and the linch pin in this coalition is the United States. The second strategic fact is that the USSR still faces threats on two fronts, China and Europe, which imposes an increasingly heavy military burden on the Soviet economic base. Third, the United States, Europe, and Japan are and will remain vulnerable to an oil cutoff or reduction and to price rises; the area of major oil supplies has become increasingly unstable and vulnerable. Fourth, that the Soviet Union will continue to enjoy major military advantages in both strategic and conventional areas for the next several years, although the trends could change later in the decade. And, finally, the Soviet Union is governed by a group of leaders in their midseventies who are not likely to shift their policies in any fundamental sense, but their successors will be caught up in some economic pressures.

The foregoing suggests that in the near term the Soviet Union will have to be active in trying to reduce the number and power of its opponents. It can no longer expect to do so through a broad negotiated settlement with the United States; this seems to have passed for the near term. Any American administration is likely to be wary of the risks of a new era of "détente"; This does not rule out contacts, negotiations, and even some agreements; but it seems to rule out a broadly based accommodation. At the same time the Soviet Union can expect to have more leverage over the course of Chinese and European NATO policy.

An examination of the Sino-Soviet relationship is beyond the scope of this chapter, but it is obvious that the Soviets intend periodically to probe for an accommodation with the Chinese and to support their political overtures with some degree of pressures and threats. The Sino-Soviet negotiations produced by the Sino-Vietnamese war in 1979 may reflect a pattern: cautious explorations by both sides interrupted by clashes over fundamental issues. After inconclusive negotiations with Moscow from September to November 1979, China broke off the talks, interpreting the invasion of Afghanistan as a major Soviet thrust for hegemony: the Chinese urged a major American response but carefully limited their own direct action, even in Pakistan. A future Chinese bid for more freedom of action was implied in the pointed announcement of an intercontinental ballistic missile test in May 1980.

A variant of this strategy is for the USSR to try to entice the Japanese into a Soviet sphere at the expense of China and the United States, probably a hopeless strategy in which the Soviets have little faith. The minimum precondition for this strategy would be the settlement of the Kurile problem, which has remained deadlocked for decades. However, the prospect of a gradual Japanese rearmament could change the Soviet view, and it may be that the Kuriles remain the ultimate bargaining chip to block Japanese rearmament.

But a more fruitful field for maneuver is the relationship with Western Europe. In the past decade the USSR has made some progress in creating a special relationship with the European NATO members. The Soviets have conceded the legitimacy of a certain level of German armaments and British and French nuclear forces. They have long since accepted the independence of the European Community, and they have also almost abandoned any enthusiasm for the prospects of Euro-Communism. And they have acquiesced in an increased level of exchanges between Eastern and Western Europe. In short, the Soviets have achieved a fairly stable relationship. The price for Moscow has been irritating but tolerable; the acceptance of a certain amount of agitation over human rights in the Helsinki framework, occasional European outbursts about aggressive Soviet activities outside Europe, and a further distancing of the Soviet Union and Western Communist parties.

Soviet strategy seems set for a new phase, generally, which will have as its critical elements the division of Europe from the United States and the political domination of Western Europe. The main tactics will be an intensification of the most recent course: to exploit a divisible détente so that the interests of the United States and the European nations gradually drift apart in major areas: in trade relations, in East-West exchanges, in political dialogues and summits, in the general climate of relations, and in the assessment of the requirements for European security and the role of negotiated arms control. It is highly unlikely that the Soviets will demand or even propose the formal retraction of American power in Europe. Much more likely is a strategy of paring down American influence with its allies, of forcing the United States to accept Soviet-inspired regional arms control agreements or arrangements in the name of Atlantic solidarity, of playing on the genuine divergencies of Europe and the United States in the Third World and thus eventually making U.S. policy a liability to Europe and an obstacle to security guarantees from the USSR.

What the Soviet Union hopes to accomplish and what it will actually realize are, of course, quite different. Soviet leaders, from Stalin through Khrushchev have launched ambitious schemes to dominate Europe, only to fail dramatically in Berlin and ultimately in Cuba. The chief difference now lies in the changing balance of power and the Soviet determination to take advantage of it. In some respects the Soviets are embarking on a subtle version of Khrushchev's Sputnik campaign: The threat of military superiority is a common theme, but the offers and demands will be much more skillfully attuned to European and even American opinion. The process is more protracted and more intricate, and no highly focused confrontation is intended (though it might develop). It is fundamentally a test of will and tenacity, however, just as the Sputnik offensive was a test of nerves.

Soviet prospects, of course, depend to a great degree on the ultimate Western responses to both pressures and conciliation. But the effectiveness of the Soviet strategy may also be influenced by events inside the USSR and within the Soviet empire in Eastern Europe.

The Soviet succession is merely a matter of time, despite the surprising durability of President Leonid Brezhnev and some of his cohorts. It is tempting to generalize about succession periods: that they are a time for change, for entertaining new policies, for repudiating the past, for a fresh start; they are times of infighting and quarrels and settling of accounts, of paralysis, of conservatism, or they are periods of historic upheaval leading to sweeping changes; and so forth.

The Soviet succession seems to be becoming both a mystery and a suspense story. It appeared in 1976 that the 25th Communist Party Congress would be Brezhnev's last. But he surprised outside observers by forcing his way into the presidency rather than retiring; and despite signs of faltering health, he emerges stronger in the hierarchy. Now one must forecast that the 26th Party Congress will surely be his last. But even now the succession has become a protracted process of change, in which the inability of the system to shift from one leader to another or from one generation to another builds up the potential for a crisis. In short, the odds for a major upheaval will grow in proportion to the delay in naming and transferring power to a clear successor.

This having been noted, however, there is ample evidence that the forces of conversatism—the party apparatus, the KGB, the military, are strongly entrenched as a result of trends in Brezhnev's policies. There is little prospect that out of the Brezhnev politburo there will emerge a genuine economic reformer, for example. Nor is it likely that the main lines of Brezhnev's foreign policies will be sharply altered. In Europe at least, there is little reason to jettison a policy that appears, on the whole, to have been successful and still holds some promise, providing the policies are skillfully manipulated (whether a new leadership will have this skill is a question). So in the short run there is no significant prospect that Soviet policy toward the alliance or toward the United States will be altered simply as a result of the change in leadership (this begs the question of whether and how Western policy might exploit a succession phase).

But this new leadership will not operate in a vacuum. It cannot escape the realities of the Soviet economy, which in the late 1980s will be in some degree of crisis. The main elements are already apparent: slow growth rates, dwindling manpower, faltering productivity, and probably an energy crisis, to say nothing of the perennial agricultural crisis. Three possibilities are suggested: first and most likely, muddling through; second, major structural reforms; and third, a major foreign crisis as a diversion. The latter option is a real danger. It is the task of Western policy to make the temptation of foreign adventure simply too risky and costly. But this raises a major Western policy question: how to respond to a shift in Soviet economic policy that would include some willingness to moderate political behavior for economic gains. And an outside possibility is a unilateral cutback in defense.

Finally, there is the ever present threat to the USSR of an Eastern European explosion. At this writing the Polish crisis is far from over, but it already appears to be far more significant that the Czech crisis of 1968. The Czech crisis proceeded within the framework of the ruling party, whereas in Poland the threat to Leninist stability arises outside of the party and its related apparatus. Moreover, there has been a unique alliance between intellectuals and workers in Poland. Although the union Solidarity's tactics have been shrewd and carefully calibrated to fall short of confrontational demands, the threat to party domination and to Soviet influence is a clear and present danger for the USSR. Independent institutions have no place in the Leninist order: they must be eliminated. Thus with each success the new Polish revolution ironically moves closer to a showdown with the Soviet Union.

Whatever the immediate outcome, the Polish crisis once again points up the dilemma for the Soviet policy in Europe. Its first priority is to protect the homeland, and this means securing a firm grip on Eastern Europe. The second priority is to extend its influence over Western Europe; currently this means a relaxation of tensions, but with continuing military advantage. If the Soviets are required to intervene in Poland or elsewhere to retain their base, then the prospects for using détente as a means of gaining increased influence over Western Europe would almost certainly evaporate for some considerable time. Thus the great irony is that after thirty-five years, the future of Soviet policy will be fought out, not in the chancelleries of Paris or London or Washington or on the battlefields of the north German plain, but in the shipyards on Danzig, where, in a historical sense, the struggle for the mastery of Europe began in 1939.

NOTES TO CHAPTER 2

1. From a memorandum by Charles Bohlen, December 1943, in *U.S. Foreign Relations, 1943, The Conference at Cairo and Teheran*, p. 845).
2. Velko Micunovic, *Moscow Diary*, p. 44.
3. See Coral Bell, *The Diplomacy of Détente*, pp. 107-108).
4. See Adam Bromke, "The Cliff's Edge," *Foreign Policy* 41 (Winter 1980-81): 159.
5. From a speech by President Brezhnev on November 2, 1977, printed in *Survival* (January-February 1978): 32.
6. *Economist* (October 6, 1979).
7. James O. Goldsborough, "Europe Cashes in on Carter's Cold War," *New York Times Magazine* (April 27, 1980): 49.
8. See *Survival* (January-February 1978).
9. March 2, 1979.
10. Press conference of June 25, 1979.
11. For an excellent analysis of the Soviet decision, see Seweryn Bialer, "A Risk Carefully Taken," *Washington Post*, January 18, 1980.
12. See Henry Kissinger, *The White House Years*, p. 946.
13. See Paul Nitze, *Foreign Affairs* (Winter 1980).

3 U.S.-ALLIED RELATIONS
The Current Crisis in Historical Perspective

Robert E. Osgood

The North Atlantic Treaty Organization is unique in the history of multinational alliances in sheer longevity. More extraordinary is its basic solidarity over a long period of constant change in the fundamental military, economic, and political conditions of its existence. But NATO is also notable for tensions and anxieties and the periodic crises among its members, which have made "disarray" a commonplace description. Paradoxically, this disarray is a byproduct of the alliance's longevity and solidarity. It reflects the continual accommodation of divergent interests among the allies within a remarkably stable East-West structure of countervailing power. Indeed, the pattern of friction within cohesion is so familiar that it becomes almost reassuring. Nevertheless, there are reasons for thinking that this time the alliance's problems are different in nature—more difficult to resolve and more dangerous if left unresolved.

The current tensions and anxieties in the alliance began with the divergences of interest and policy between the United States and its major allies that followed the prolonged capture of American hostages in Iran and the invasion of Afghanistan. These divergences caused a widespread if ill-defined perception in the United States and abroad that the alliance is in a state of crisis that goes beyond chronic disarray to impinge on its cohesion and security. This chapter explores the sources of this feeling and assesses their consequences. It concludes that the alliance face a serious crisis, although it may remain deceptively latent or quiescent until some new catalyst activates it—unless the allies can collectively neutralize the ingredients of the crisis. The analysis starts by distinguishing between some fundamental sources of tension that have underlain allied relations, particularly relations between the United States and its European allies, almost from the beginning of developments that troubled these relations in the 1970s.

Tensions between the United States and its European allies are especially significant because the American commitment to regard the defense of Western Europe as seriously as the defense of the United States has been indispensable to the security of the allies and to their cohesion. It has been crucial to the accommodation of Germany's resurgent strength. Although at the time this commitment was undertaken few sought or foresaw creation of the integrated military organization, NATO, that became identified with it in the aftermath of the Korean War, the organization itself and the military arrangements undertaken to support it, including the U.S. military presence in Europe, came to be regarded as the keystone of the alliance. This transformation of the alliance from a guarantee pact to a military organization has greatly strengthened the alliance, but it has also made it more susceptible to tensions arising from certain fundamental conditions and developments in the first 20 years of its existence.

- The security of the European allies basically depends on the United States because of the dominant role of U.S. nuclear deterrence in containing the USSR. Notwithstanding West Germany's preponderant role in forward defense and the development by the United Kingdom and France of independent nuclear forces, the allies have remained to a considerable extent military dependents of the United States as Soviet military strength has increased across the board and American troops in Europe have come to be regarded as the indispensable guarantee of the American deterrent. This position of continuing dependence—the psychology as much as the fact of dependence—was bound to be the source of mutual dissatisfaction and anxiety as the European allies recovered their strength and individuality, as U.S. and European views on East–West relations and other matters diverged, and as Soviet nuclear capabilities called into question the credibility of U.S. extended deterrence.

- The political necessity of NATO's commitment to defend the Federal Republic of Germany on the forward line but the great difficulty, if not practical infeasibility, of doing so with conventional forces alone (for a combination of geographic, strategic, and politico-economic reasons) pose two insoluble problems, which have been exacerbated by the growth of Soviet nuclear capabilities against the United States and Western Europe: how to defend Western European countries by nuclear weapons without destroying them and how to maintain a credible nuclear deterrent against conventional attacks when using it seems likely to destroy the European allies and devastate the United States. These problems, though insoluble, have been alleviated by adoption of a strategy of flexible and controlled response (both conventional and nuclear) and by the seeming implausibility of an actual Soviet attack. But the continued growth of the Soviet Union's nuclear striking power, coupled with its inherent military advantages in Central Europe, keeps alive the psychological strain on U.S.-allied relations.

- The United States approaches East-West relations from the standpoint of global containment; the allies, more from the standpoint of regional or local interests. As the geographical scope of U.S. security interests has expanded, the scope of its allies security interests has contracted, thereby accentuating the difference of perspective. This difference of perspective is particularly divisive when the United States incurs military risks or takes military actions that its allies fear will threaten their own security or the benefits of détente or when the United States seeks accommodations with the USSR that the allies fear will sacrifice their regional or national interests.

- The United States not only has global concerns and interests that are not shared equally or at all by its allies. It also enjoys a greater insulation from the pressure of Soviet sticks and carrots, both military and economic, and is therefore under less need or inducement to conciliate Moscow; whereas European efforts at conciliation, the U.S. fears, may weaken the utility of American sticks and carrots against the Soviet Union without moderating Soviet behavior.

- The steady spread of the cold war throughout the Third World, accompanied by the rising instability of governments and conflict among states in those heterogeneous nations, have been abundant sources of local wars and crises in which the interests of the United States and its European allies may diverge, especially when American involvements have diverted U.S. resources from the protection of Europe or have threatened to involve European countries in a larger East-West conflict.

- As the European allies have become economically thriving and politically self-confident industrial giants, they have also become more determined to assert their political independence from the United States in trade, monetary, and diplomatic relations. But they have become no less anxious to retain American military protection and no less reluctant to take over the entire burden of their own defense. This division of labor suits both sides of the Atlantic well in spite of the friction it occasionally engenders. But coupled with increased policy divergences, the problems of equitable burden-sharing, or any serious prospect of a major U.S. retrenchment in Europe, it can be the source of resentments and fears that could undermine the "transatlantic bargain" (in Harlan Cleveland's felicitous phrase).

The following basic conditions have repeatedly caused friction between the United States and its allies with respect to a variety of issues: the rearmament of West Germany, military strategic doctrine, defense burden-sharing, national nuclear forces, trade and monetary relations, U.S. intervention in conflicts outside the NATO area, the Arab-Israeli conflict and its resolution, the Palestinian issue, East-West negotiations on arms control and European security, and U.S. human rights concerns. Sometimes allied differences have reached the level of crises, as during the Suez War, President De Gaulle's challenge to American "hegemony," the U.S. rejection of the Skybolt missile for the British, the proposal for a multilateral nuclear force (MLF), the Yom Kippur war, and the Greek-Turkish dispute over Cyprus.

None of these issues or crises, however, has called into question the basic purpose and structure of the North Atlantic alliance. None has come close to undermining its essential cohesion and security. In each case a readjustment of allied policies and relations has restored relative harmony and cleared the way to cope with the underlying issue. This will probably continue to be the case, but because of convergence of some major international developments in the 1970s an adjustment of policies and relations is needed that is more substantial and more difficult than before.

Added to these conditions, which have underlain recurrent disarray from the beginning, additional developments now pose an unprecedented challenge to the cohesion and security of NATO:

• The Soviet Union, since the time of the Cuban missile crisis, has invested steadily and massively (at the rate of about 4 percent annual increase) in the growth of its armed forces in every segment of military power, while the United States—especially in the 1970s, until Congressional pressure against SALT and the Soviet invasion of Afghanistan reversed the trend—reduced its investment. It did so partly in the expectation that the Soviets would settle for strategic parity or "essential equivalence," which would be embodied in an agreement on strategic arms limitations, and partly in response to the Vietnam syndrome. As a result the USSR has added to its longstanding superiority in conventional forces in the European theater, buttressed by the modernization of Warsaw Pact forces, a superiority in theater nuclear forces and a forthcoming superiority (by the mid-1980s) in intercontinental nuclear forces. This novel situation heightens Western Europe's, and especially Germany's, sense of physical vulnerability, intensifies allied anxieties about the reliability and effectiveness of U.S. protection, and enhances the Soviet ability to decouple Soviet-European détente from U.S.-Soviet tension.

• The development of a U.S.-Soviet détente, of which SALT was the centerpiece, was accompanied by the development of a European détente, which was based on the four-power Berlin agreement and complementary agreements between the two Germanies and other bilateral Eastern agreements, the Conference for Security and Cooperation in Europe, and the Helsinki Final Act. These European treaties essentially recognizing, in order to stabilize, the East-West division of Europe, were given an economic and human dimension through East-West trade agreements, the importation by Germany of natural gas from the USSR in exchange for large-diameter gas pipe, some 8 million West Germans annually visiting East Germany, and more than 60,000 Germans annually emigrating to the West from Eastern Europe. This European détente is regarded by America's allies as a crucial counterpart to collective defense in an overall security strategy. It has given the allies, especially Germany, a tangible stake in conciliating the Soviet Union and decoupling their relations with the Soviets from tensions and crises between the U.S.

and the USSR, particularly when these conflicts spring from issues outside the NATO area. The United States, on the other hand, being more directly concerned with checking Soviet global advances, having less of a tangible stake in détente, and being less vulnerable to Soviet sticks or carrots, is inclined to view East–West relations as indivisible and to deplore a differentiated East–West posture that protects allied fruits of conciliation at the expense of a common front against the USSR. The Soviet Union, Washington fears, tries to exploit this difference of perspective and interest by decoupling European from American relations with the USSR so as to divide the alliance and offset American opposition to Soviet moves. The Soviet invasion of Afghanistan brought this difference of attitude toward détente to the forefront of U.S.–European relations, as the Carter administration suspended the process of ratifying SALT and took other measures to reinforce containment militarily, diplomatically, and economically. But only a Soviet invasion of Poland and protracted fighting might have a comparable effect on the European détente.

• Contrary to some American hopes and expectations in the aftermath of the war in Vietnam, the Third World has been beset with widespread political instability, revolutionary movements, and national and communal conflicts. This turmoil has provided not only new opportunities for the projection of Soviet influence and presence but also for indigenous threats to Western access to oil and other vital resources.

• Soviet capacity to project armed force into the Third World through arms aid, Cuban troops, East German technicians, and its own naval, air, and ground forces has greatly increased in the last decade. At the same time, U.S. projection forces and will to use them have atrophied in the post–Vietnam atmosphere, and the political conditions of maintaining and deploying them in troubled areas have grown less hospitable and more complicated. With the help of its new capacities the Soviet Union has proved to be eager and able to project its presence and influence. Its intervention in Ethiopia, its dominant position in South Yemen, the collapse of Iran, its concentration of forces on Iran's northern flank, and the invasion of Afghanistan now bring Soviet power to a focus on the Greater Middle East (Southwest Asia, the Persian Gulf and Arabian Peninsula, and the Horn of Africa). This area is intrinsically far more important to the security of the United States and its European allies (and to Japan) than any previous point of Soviet pressure in the Third World. The demonstrated difficulty the USSR has in maintaining clients and positions of influence beyond the reach of the Soviet army's occupation over a long period may continue to hamper Soviet efforts to gain parity with the United States as a global power, but, given Soviet persistence, this is not much consolation.

• In the 1970s the United States and its allies (with the partial exception of the United Kingdom and Norway) greatly increased the dependence of their economies on the importation of Middle Eastern oil, so that now one-Third of U.S. imports and two-thirds of West European imports come from the Persian Gulf. Consequently all the allies have acquired a vital security interest in the unobstructed production and shipment of oil at tolerable prices. This interest is now threatened by the convergence of several developments: political instabilities and international turmoil in the Middle East, OPEC's fourfold increase in oil prices, the OPEC members' exploitation of Western (and Japanese) oil dependence for foreign policy objectives, and the growing Soviet presence and ambitions in the area. Ideally this magnification of a common security threat to the United States and its allies should have galvanized them to take concerted countermeasures. In reality it has accentuated differences of interest and perspectives for several reasons: The greater dependence of the European allies on Middle Eastern oil is related to differences of policy with respect to the Arabs, Israel, and the Palestinians, which reflect the American special relationship with Israel and its role as peacebroker as well as the European anxiety to accommodate the major oil suppliers. The mounting prospect of competition between the United States and its allies for access to a growing shortage of oil supplies in relation to demand, as reserve-to-production rates decrease and exporting countries cut production portends further friction between the United States and its allies. The European allies' emphasis on the need for nuclear energy to help supplant oil conflicts, which they believe depends on reactors producing weapons grade material, has conflicted with the commitment of the United States to the necessity of nuclear nonproliferation, which called for proscribing the production of weapons grade material in reactors or strictly accounting for it through international safeguards. The great disparity of military role and influence in the Persian Gulf region between the United States and its allies puts the United States in the position of assuming the burden and risks of containment on behalf of allies that fear it may jeopardize détente and leads the allies to look for ways to insulate themselves from regional conflicts.

- The Soviet Union, according to many official and private estimates, may become a net importer of oil from the Middle East as early as the mid-1980s. Soviet entrance into the world market for oil, it can be argued, may give Moscow a new interest in the stability of regimes and the moderation of international conflicts in the Middle East. But it seems more likely, given the existing political instability and international turmoil and the opportunities they offer for projecting Soviet influence, that the Soviets will seek control of oil production, directly or indirectly (that is, through Moscow-oriented regimes). Not only could they thereby secure a reliable oil supply; they might also manipulate the terms under which energy will be supplied to the United States and its allies so as to exert leverage on the allies for Soviet arms control and other policies that diverge from those of the United States.

- In the 1970s the combination of growing economic interdependence and the deterioration of the Bretton Woods international economic system—marked by the decline of America's competitive trade advantage, the weakening of the dollar as the international reserve currency, and the breakdown of the exchange rate system—generated chronic tensions concerning trade policies and monetary relations between the United States and its European allies (and Japan). The fourfold increase of oil prices following the 1973 Arab-Israeli war and the onset of persistent stagflation in all the industrial democracies aggravated these tensions and gravely weakened the whole international economic underpinning of Western security and, with it, America's bargaining power with its allies on defense burden-sharing, as well as on trade and monetary issues.

- The marked erosion of American economic and military primacy in the last decade, partly for reasons virtually beyond U.S. control (such as the growth of Soviet military strength, the economic resurgence of America's allies, and the sudden rise in oil prices) and partly because of the domestic and international effects of the traumatic Vietnam/Watergate era, has undermined America's reputation for maintaining and effectively managing power in the world. The Carter administration, registering all too accurately the nation's tension between hopes for a new and morally satisfying international order and its rediscovery of the cold war, accentuated the impression of declining power by its penchant for vacillation and contradiction. This impression and the element of reality behind it encourages Soviet penetration of the Third World, intimidates weak and insecure states vulnerable to this penetration, discourages Third World countries from undertaking security arrangements with the United States, impedes cooperation between the United States and its allies with respect to joint security problems outside the NATO area, and has tended to distract attention from the objective of containing Soviet power by focusing anxieties upon American imprudence and ineptitude.

These foregoing developments constitute a fundamental change in the comparatively benign international conditions that underlay the great creative period of alliance growth and consolidation from 1949 to the late 1960s. The tensions and crises of that period seem in retrospect like the growing pains of robust youth; those since the early 1970s seem more like the adversities of maturity. Nevertheless the allies have approached some of these adversities with a reasonably constructive and cooperative adjustment of policies and institutions, which has at least saved them from senility. Two areas of adversity and adjustment are particularly important: trade and monetary relations and harmonizing détente with defense.

TRADE AND MONETARY RELATIONS

In the 1970s America's fading technological and industrial prowess and its declining competitive advantage in exports while the U.S. economy became more export-dependent, not to mention the effects of the Vietnam War and inflation, exerted a crippling effect on the value of the dollar and its utility as a reserve currency and undermined the viability of the Bretton Woods system of foreign exchange. Devaluation of the dollar, realignment of currencies within wider bands of fluctuation (the "tunnel"), European efforts to narrow the band of fluctuation through a joint float of European Community (EC) currencies (the "snake"), the replacement of the tunnel with a system of managed floating exchange rates, and more recently the creation of a European Monetary System (EMS) have failed to stabilize the value of the dollar; but the dollar remains the only acceptable international intervention currency. The deutsche mark, although the strongest currency in the floating system and a powerful influence in protecting the value of the dollar as well as affecting the value of European currencies, cannot become a substitute for the dollar. The German economy is not large enough or strong enough to support this role. The German government fears that making the deutsche mark a major reserve currency would jeopardize well-established domestic policies (for example, the control of inflation and the preservation of a favorable competitive position in export markets) and shatter the EMS, which was fashioned by German Chancellor Helmut Schmidt and former French President Valéry Giscard d'Estaing in order to protect EC currencies and trading ties against fluctuations of the dollar and instability in the international trading system.

To these monetary difficulties are added the unresolved problems of economic security that spring from rising oil prices, the possibility of an economically disastrous prolonged blockage of Middle East oil supplies, growing European competition with the United States for Third World markets, especially in high technology, where the United States has lost its dominance, and ominous protectionist tendencies.

Nevertheless, through a constellation of economic organizations and a succession of summit meetings, as well as a number of adjustments between West Germany and the United States, the Atlantic allies (and Japan) have painfully achieved compromises that have avoided a breakdown of the monetary system, ameliorated international differences and tensions arising from trade and monetary conflicts (especially U.S.-EEC and Franco-German), and kept alive the disposition of all allies to reach pragmatic compromises in recognition that their economies are fatefully entangled and critically dependent on preserving a relatively liberal, open trading system.

HARMONIZING DÉTENTE WITH DEFENSE IN EUROPE

The 1970s posed new problems of managing East-West relations. On the one hand, the development of détente put agreements on the limitation of arms, new rules of East-West relations, and the development of human contacts and cultural and commercial exchanges at the top of the agenda of Western foreign policies. This was especially true in the Federal Republic of Germany, which came to view the new relationship with the Soviet Union and the Eastern bloc as indispensable to national security in a divided Europe. In the West Germany and the smaller NATO countries, moreover, there arose a burgeoning antinuclear antimilitary movement. On the other hand, the modernization of Warsaw Pact forces and the steady advance of Soviet military capacity across the board, in relation to Western capacity eventually put a premium on stopping a relative military decline, which if permitted to continue might not only jeopardize NATO's defense posture but undermine the military equilibrium on which détente is based. Although theoretically détente and defense should complement rather than oppose each other, in reality the political psychology at work in democratic countries makes it difficult to increase defense efforts while sustaining efforts toward East-West accommodation. This difficulty is compounded by the development of differential U.S. and European benefits from détente and differential sensitivities to the expansion of Soviet influence outside the NATO area.

Nevertheless allied governments have responded, if belatedly and perhaps too weakly, to the necessity of maintaining the military balance in Europe through additional defense expenditures while maintaining a more or less coordinated policy on arms control and other diplomatic relations with the Soviet Union. In some cases the need to coordinate arms control positions and tactics has led to extraordinarily effective collaboration at the working staff level—for example, the work on MBFR by the special consultative group on arms control. Although strong liberal and left wing constituents in some European countries make defense efforts contingent on arms control efforts, it is hard to demonstrate that the latter have seriously interfered with the former (which is different from arguing that defense efforts have been adequate). The proposition endorsed by NATO members in the Harmel report of 1967, that détente and defense should proceed in parallel, states a fact of political life that NATO governments have so far used more successfully to advance defense programs than actual arms control agreements. Thus the agreement of December 1979 to deploy a modernized long-range theater nuclear force (LRTNF) of 572 Pershing IIs and cruise missiles was won without disabling concessions to Soviet efforts to substitute a one-sided arms control agreement. The allies agreed only that deployment of the new Europe-based weapons should be contingent on arms limitation talks with the Soviet Union (and, to satisfy Germany, contingent on at least one other continental NATO country emplacing the new missiles on its soil.

Contrary to Henry Kissinger's initial consternation, Germany's pursuit of *Ostpolitik*, which was greatly facilitated by Kissinger's pursuit of SALT, did not touch off a rush to Moscow or a relapse into neutralism. Rather, it resulted in parallel initiatives and a good deal of policy coordination between the United States and Germany, which produced a set of treaties vital to European security. Although the Conference for Security and Cooperation in Europe (CSCE) occasioned considerable worry that the Soviet Union would extract a propaganda victory and exploit it to consolidate and legitimate its sphere of control in Eastern Europe, CSCE has turned out to be at least as advantageous to European countries, both East and West, as to Moscow and has not been a source of major substantive differences between the United States and its allies.

As with trade and monetary issues, the substantive results of the process of consultation, compromise, and collaboration in carrying defense and détente forward on parallel tracks have fallen far short of solving the problems to which they are addressed. The bright expectations for arms control as an instrument of national security and détente have been dimmed by the suspension of SALT and the stalemate of MBFR, while the United States and its allies allowed the military balance to deteriorate to an extent that the effort to restore the balance can be represented by Moscow as a provocative change in the status quo.

This situation produces a growing divergence of views between Americans and Europeans on the utility of arms control. While European governments and their constituencies desperately cling to the process of arms control as something indispensable to peace and the avoidance of an unlimited arms race, in the United States basic assumptions underlying the whole arms control movement—for example, that agreed arms limitations promote a more stable, safer, and cheaper military balance—are undergoing revision on the basis of the experience of the 1970s. Even the antiballistic missile (ABM) treaty, which along with the nuclear test ban was widely viewed as the outstanding success among arms control efforts, is subject to critical reappraisal as the shift in the military balance and technological developments of the last decade make hard-point ballistic missile defense a more attractive means of reducing the vulnerability of ICBMs than when the ABM treaty was signed.

New and more formidable problems of arms control lie ahead. Even if the effort to redress the East–West military balance through the deployment of 572 medium-range Pershing IIs and cruise missiles in Western Europe survives Soviet arms control appeals directed at Western European publics and parliaments, the active renewal of arms control negotiations will encounter extraordinarily difficult political and technical problems. Negotiations must deal with so-called Eurostrategic weapons, including forward based systems (FBS) and perhaps allied nuclear forces; with conventional theater weapons, now the subject of MBFR; and with intercontinental nuclear weapons, for these systems are more than ever interrelated; yet the experience with SALT shows the great difficulty of formulating a mutually advantageous comprehensive arms limitation agreement. Moreover, as European governments confront the concrete issues of the arms control process, such as how to handle FBS, troublesome differences of interest and perspective between them and the United States will emerge. If, in the midst of unproductive arms control negotiations, economic stringencies should induce allied governments to renege on their commitments to long-term defense programs while the United States is launched on a major defense buildup, there would be great strains on the whole bargain to conduct arms control and defense in parallel.

The creditable record, so far, of dealing with the problems of the Atlantic economy and combining defense with détente is no assurance that the allies will cope successfully with the new problems posed by security threats in the Third World, beyond the area of NATO commitments. Involvements of the United States and some of the European allies in crises outside the alliance area have occasioned considerable tension and anxiety among the allies before. One remembers, particularly, the Suez war, the war in Vietnam, the civil war in Algeria, and the Yom Kippur war. For obvious historical and geographical reasons there has seldom been a consensus between the United States and its allies on how to deal with local wars and crises in the Third World. But in the present convergence in the Greater Middle East of some disturbing trends of the 1970s, there are the ingredients of a crisis substantially more serious and more difficult to manage than any of those in the past.

That recent developments in the Middle East constitute such a serious crisis is not obvious on the face of things. Toward the end of the Carter administration there arose an extraordinary intuition in Europe of a crisis in U.S.-European relations, turning upon apprehensions of American militancy combined with uneasiness about American impotence. This feeling has been aggravated by the hard-nosed and critical official American approach toward allied governments, which also reflects a feeling of crisis in the United States—but a feeling directed first against the Soviet danger and second, against the perceived failure of the Europeans to respond to this danger. Nevertheless, if one measures this crisis by the intensity of clear disagreements between the United States and its allies on specific issues, it is relatively benign. The reason is that there is not much of a concrete nature to disagree about as long as little is being done. The U.S. government did not ask its allies for substantial—that is, really costly—as opposed to largely symbolic sanctions; and after the aborted effort to rescue the American hostages in Iran, it refrained from openly contemplating military action. The allies successfully resisted most of the requested sanctions, asserted their diplomatic independence with respect to the Arab-Israeli peace issue and the lines of communication with Moscow, and voiced critical apprehensions about the prudence of American actions; but they made enough concessions to American pressure to preserve the facade of solidarity.

What makes this crisis qualitatively different and more serious as compared to previous crises are the following conditions:

- Never before have local instabilities and Soviet capabilities and behavior in the Third World threatened both U.S. and Western European security interests so vitally. The Soviet willingness and ability to use its own armed forces to dominate a country outside the postwar sphere of control established by the Red Army, combined with the Soviet positions in South Yemen and Ethiopia, and massive Soviet forces on the Iranian border, and the proximity of Afghanistanian air bases to the Persian Gulf, and ominous developments in a period of history in which Soviet leaders seem determined to break out of Russian "encirclement" and use their new military parity as the foundation for establishing global parity. Added to these international developments are equally disturbing developments indigenous to the area: the collapse of Iran, the only major regional military counterpoise to Soviet military power; the fragile internal stability of Middle Eastern countries; the burgeoning national, ideological, and religious differences among them; the Arab-Israeli-Palestinian conflict; the Iran-Iraq war. All these conditions, together with the extreme dependence of the United States and its allies on Middle Eastern oil and the obscure but possibly dangerous effects of the projected Soviet importation of Middle Eastern oil in the 1980s, constitute an unprecedented threat to the NATO alliance itself. The threat is no less serious because it could take a number of unpredictable forms and because it could be precipitated by the actions of, and developments within, countries that are beyond the control of the superpowers. One need not postulate anything so extreme as the complete blockage of oil supplies, a Soviet military invasion, or the involvement of the United States and the USSR in a proxy war in order to appreciate the magnitude of this threat.

- This security threat occurs not only outside the area of NATO commitments but also in an area in which the United States and some of its European allies have considerable differences of national economic and political interests (for example, on the Palestinian issue and relations with Arab countries), together with a substantial difference in vulnerability to potential restrictions on oil supply and an uncertain system of interallied reallocation of supplies in the event of a serious restriction. These differences accentuate the conflict between regional and global perspectives between the United States and its European allies and compound the difficulty of addressing the common security threat in concert.

- The Soviet threat in the Middle East raises questions about the management of East-West relations, particularly with respect to the divisibility of détente, which go to the heart of the commitment to pursue détente and containment simultaneously and which are potentially very divisive between the United States and its allies. To Americans, the European insistence on preserving the achievements of the European détente despite Soviet actions and regional conflicts that threaten European security even more intensely than American security seems like a typical manifestation of allied irresponsibility. To Europeans, American efforts to impose collective sanctions on the USSR or the possibility of American military deployments and actions in the Persian Gulf portend a willingness to sacrifice the achievements of European détente for the sake of an ill-considered and dangerous posture of confrontation that will jeopardize Western security—which looks to them like another demonstration of American immaturity.

- The problem of organizing an effective concerted response to this multifaceted military, economic, and political crisis is compounded by the loss of America's primacy in the international economic system and the central East–West military balance and by the decline of America's reputation for the competent, prudent management of power and diplomacy. This decline of reputation and therefore of confidence began with the war in Vietnam. It was deepened by the trauma in American society and the affirmation of fragmented Congressional power against the executive control of foreign affairs following the war. It was consolidated by certain elements of style and substance in the Carter administration, particularly its disposition to vacillate and convey contradictory signals and its pursuit of human rights and nuclear nonproliferation policies that seemed to jeopardize détente and national energy needs (not to mention the export market for nuclear reactors).

For these four reasons, developments that pose potentially the most serious threat to the security of all the allies, instead of galvanizing the allies into a common response, tend to divide them and impede effective countermeasures. Perhaps the crisis must get demonstrably worse before the allies begin to work together to cope with it.

It takes little imagination to see how what is now largely a latent crisis could become an active one, given any number of possible catalysts. The Iran–Iraq war is obviously one such catalyst—indeed, it is hard to foresee an outcome that will be anything but disadvantageous and dangerous—but there may be others: an overthrow of the Saudi regime by radical forces, the detachment of Baluchistan with Soviet support, the resurgence of the move toward an independent Kurdistan or Arabistan with Iraqi and Soviet support, the unification of a Marxist, pro-Soviet Yemen, the renewal of revolution in the Dhofar region of Oman, the death of Sadat, a relapse into anarchy in Turkey.

One does not have to assume rashness in Soviet behavior to appreciate the danger that Moscow's enhanced interest in reliable access and influence in this troubled area, coupled with its increased oppotunities to remove Western influence from an area of historic ambitions and growing geopolitical significance, now under the shadow of a superior concentration of Soviet military power along the northern border, could give any of these catalysts the gravest global significance. President Carter's doctrine declaring that the U.S. regards "an attempt by any outside force to gain control of the Persian Gulf region" as an "assault on its vital interest," which "will be repelled by any means necessary, including military force," must reinforce Soviet caution against getting into a situation that might lead to a direct encounter with U.S. forces, no matter what the local imbalance of U.S. and Soviet forces in the area may be. But there are many possible challenges to the Carter doctrine short of such an encounter. For political reasons those springing from the actions of "inside forces" that might threaten the security of friendly states and the supply of oil or invite the extension of Soviet arms and bases would be particularly difficult to respond to effectively. Moreover, no response to the actions of regional powers can be taken without anticipating the reaction of the Soviet Union.

More likely as a sudden East-West confrontation in the Persian Gulf area and probably more divisive in the Atlantic alliance would be the steady erosion of American influence in the face of superior strength, the disappearance of all effective local counterpoises to Soviet strength, their replacement by frightened or hostile regimes that feel compelled to come to terms with the USSR, the intensification of competition between the United States and its allies for access to shrinking supplies of oil, and the accentuation of interallied differences on how to deal with oil-rich Arab countries, Israel, and the Palestinians. Under these conditions Moscow's best strategy would be to console and entice the European allies with trade and arms control deals, to isolate Western Europe as an island of détente, and represent itself to the allies and others as the guarantor of regional order, peaceful access to oil, and justice for the Palestinians against the waning and unreliable American hegemony. Such a strategy might enable Moscow to establish the kind of regional preponderance it has long been prevented from achieving. It would, at the least, exert an insidious influence on the cohesion of the alliance and the economic security of the allies.

In the Middle East the ingredients now exist for either a sudden or a creeping crisis that will pose a greater threat to the cohesion and security of the NATO alliance than any previous crisis. What should the United States and its allies do to prevent, ameliorate, or nullify this crisis?

When faced with such a multiplicity of interacting factors on the one hand and such basically divisive factors within the alliance on the other, one may long for a concerted grand strategy. Such a strategy would establish an agreed division of labor among the allies within an integrated military, economic, and diplomatic plan aimed at constructing an effective regional Western military counterpoise, instituting an effective reallocation system for oil-supply emergencies and beginning to solve the problem of Western (and Japanese) oil dependence through the concerted development of oil substitutes. It would restore a stable regional equilibrium compatible with both a broader Arab-Israeli peace settlement and the establishment of reliable bases for the infrastructure and forward forces of a rapid deployment force that will not have to draw from NATO forces.

Unfortunately, this is probably a counsel of perfection. Some of the objectives of such a strategy, particularly, those pertaining to a congenial regional equilibrium, are simply beyond the power of any group of outside states to achieve. Some raise policy choices on which no allied consensus exists or is likely to emerge. All of them, together, require a level of economic and military effort, a degree of domestic support, and an extent of interallied collaboration that are beyond the realm of political feasibility.

Nevertheless the vision of a concerted grand strategy may serve as a useful model as we go about the hard business of establishing, in an inevitably piecemeal and sporadic fashion but with as much real consultation and cooperation among allies as possible, the components of an aggregate effort. Where domestic energy policies, the politics of oil dependence, differing sensitivities to Arab and Israeli views, special interests in the Third World, and other factors make allied collaboration impossible, continual consultation on an aggregate effort might at least save the allies from working at cross purposes. To this end the following outline of the principal components of such an effort is presented.

Considering recent precedents in the conduct of policy, one hesitates to concur with Zbigniew Brzezinski's ill-fated recommendation of more "architecture" for U.S. foreign policy. Nevertheless it may help to understand the limits as well as the goal of an aggregate allied effort to distinguish elements of its foundation, its general design, and the building blocks to implement the design.

The Foundation

Despite the loss of American economic and military primacy and the growing assertion by allies of national interests independent of American policies, the United States remains the most powerful member of the NATO alliance, upon which the security of its allies absolutely depends. No substantial initiatives for dealing with the present crisis can be taken without American leadership. Therefore, indispensable to an aggregate effort to cope with the present crisis is *the restoration of the allies' confidence in the efficacy and prudence of American military power.*

In the Middle East this requires the development of the rapid deployment force (RDF) as a quick-reacting, highly mobile force, supported by adequate logistics and a regional structure of facilities (ideally, with some forward ground forces) that can protect oil fields and shipments against hostile local forces and deter Soviet military intervention without weakening the forces necessary to defend NATO's central front. The size, composition, basing, deployment, and infrastructure of the RDF are controversial. Many Europeans view a six-division force as too small to cope with Soviet forces and too large and politically unsettling for the contingencies in which a military response might be useful. But American officials who have carefully assessed the military and political implications conclude that any lesser force would deprive the United States of necessary options, including an effective interposition against Soviet military moves.

If a force of this magnitude needs to be created, and if the requisite bases and facilities in the area can be secured, then the United States needs the material collaboration of its allies, not just their stamp of approval. It is all the more important, therefore, that the United States organize with its allies a process of continual consultation on contingencies, responses, and joint contributions that can contribute to the aggregate effort directly and in the NATO area. Then, even if there is not a complete consensus with respect to the use or nonuse of force in all circumstances, there will at least be a basis for confidence in the deliberated and prudent use of American power in the area.

Restoring the military balance in the NATO area is no less important than in the Persian Gulf area. It would be a serious mistake to prepare for military contingencies in the Middle East at the expense of meeting minimum defense needs in NATO, since the two areas are parts of a single strategic field. To be sure, a military encounter in the center of Europe or even on the flanks is far less likely than in the Gulf area. But not only are the stakes even greater in Western Europe; an adequate military balance in NATO is the crux of cohesion of the United States and its allies and the indispensable basis for pursuing a concerted strategy of East-West relations. Moreover, a course of action in the Middle East that tangibly weakened the U.S. commitment to the NATO area would jeopardize the prospect of enhanced cooperation with the European allies in the Middle East.

Also essential to the foundation of the aggregate allied effort to cope with the current crisis is strengthening the base of international economic cooperation on which military and diplomatic collaboration depends. This is not the place to deal with a subject that requires a book in itself. But it is worth noting that at the foundation of all foreign programs and particularly of one so entangled with oil imports and energy policies there must be satisfactory growth rates, moderate inflation, and high rates of employment in the major industrial countries in order to make macroeconomic coordination between countries feasible and microeconomic adjustment within countries acceptable.

The Design

A coherent allied effort to deal with the current crisis must be consistent with an agreed strategy for managing East-West relations. The greatest single security problem confronting the West is the expansion of Soviet influence and control in the Middle East and other parts of the Third World. To be sure, this expansion also poses new problems for the Soviet Union, carries with it new constraints as well as opportunities, and springs, in part, from a sense of encirclement and geopolitical insecurity. One may also concede that Soviet opportunities spring initially from indigenous developments beyond its control and that the volatility of internal and external politics in the Third World, along with the crudeness of Soviet imperial behavior, often upset Soviet positions of influence. But these facts do not essentially mitigate, although they complicate, the problem of containing Soviet expansion.

Faced with this problem, the West might logically decide to deal with it by sharing with the Soviets the responsibility for regional order, or it might look toward a tacit agreement on spheres of interest and influence. In reality, however, neither of these classic modes of accommodation seems feasible given the fundamental nature of conflicting East–West objectives and the volatility of the regional disequilibrium; and neither would elicit the cooperation or acquiescence of the regional states. A modicum of order therefore depends on a continuing and rather fluid contest for influence, constrained by a common recognition of the dangers of overcommitment, including the risk of a proxy war.

If a fairly intense contest for influence continues in the volatile arena of the Middle East, the issue commonly posed as the divisibility of détente may become even more divisive between the United States and its allies. The issue might be easier to resolve if its meaning were clearer. Americans who argue for the indivisibility of détente may mean three somewhat different things: (1) The allies should not permit the Soviet to enjoy the economic and political benefits of détente if Moscow pursues an aggressive course of action that threatens Western security interests in the Middle East or elsewhere. (2) The allies should not permit themselves to fall into the Soviet trap of isolating Western Europe as an island of détente, with which Moscow cultivates special economic deals and arms control arrangements while following a policy of confrontation with the United States. (3) It is politically unrealistic to continue relationships that are intended to reduce East–West tensions when East–West tensions are intensified by threatening Soviet actions outside Europe.

The first proposition makes sense in terms of cold *Realpolitik* only to the extent that the allies, by making the benefits of détente contingent upon Soviet good behavior, would gain more than they would lose or to the extent that raising the cost of Soviet misbehavior at the expense of détente can serve as an effective warning, deterrent, or constraint. It is hard to see how putting at risk or terminating the network of agreements that grew out of *Ostpolitik* could achieve any such advantage. They are very much to the advantage of the West regardless of East–West conflicts over Middle Eastern or other issues. The European allies are understandably reluctant to give up the benefits they derive from these agreements for no tangible advantage.

The same logic should apply to arms control. Thus if we reject a particular SALT negotiation or agreement, we should do so because the terms are flawed (which is plausible), or because we hope to get a more advantageous agreement (which is more likely if we redress the strategic balance), or because we can use the option of rejecting SALT as a lever to restrain Soviet behavior (which is quite unlikely), but not because we wish to punish the Soviets by depriving them of a favor or simply because we want to demonstrate our anger. The arms control process has already come to carry too much extraneous political baggage, as though it were indispensable to détente almost regardless of the actual prospect of reaching an agreement that serves Western security. The United States will not promote a more realistic approach to arms control if it pursues the counterpart of this approach by rejecting the process simply because the atmosphere of détente breaks down on other issues.

Whether the imposition of economic sanctions can gain more in terms of moderating Soviet behavior or weakening Soviet capabilities to misbehave than they cost is a more complicated question, which is dealt with in the discussion of building blocks.

The second proposition, which aims to prevent Moscow from dividing the U.S. from its allies, is compelling. This kind of division of East–West relations could only benefit the Soviets while weakening the alliance.

The third proposition is certainly true to some extent. Obviously, sheer outrage against the grosser forms of Soviet behavior must have an impact on arms control negotiations, trade agreements, and other relationships that might otherwise be of mutual interest strictly on grounds of *Realpolitik*. Trouble arises when American outrage is not shared by European governments and citizens, as in the case of Afghanistan (though not, one can safely predict, in the case of a Soviet invasion of Poland). If only because of the problems of differential outrage and the melancholy fact that outrage (as after the invasion of Czechoslovakia) sooner or later runs its course, it is important to distinguish, analytically, between concessions to public sentiment or affirmations of basic principles of international decency and the requirements of statecraft.

The counterpart of an agreed allied strategy of East–West relations must be an agreed view of the basic structure of relations between the United States and its allies. In this respect the issues raised by the Middle East crisis pertain to the distribution of military, economic, and diplomatic roles and efforts among the allies for the sake of common security interests.

From a standpoint persuasive to many Americans, the United States cannot be expected to assume a greatly augmented effort to protect common interests that are even more vital to its allies unless the allies make visible, commensurate contributions. Materially the United States needs complementary efforts by allies, proportionate to their resources. Politically it needs multinational cooperation in order to deal effectively with regional powers and the Soviet Union and to demonstrate to Congress and the American public equitable sharing of burdens and risks. From an American global perspective, that the major West European allies should continue to confine their security efforts so largely to their own region when they have achieved the status of economic superpowers and their very economic survival is at stake in a crisis in the adjoining region tends more and more to look like an unacceptable anomaly.

From the European standpoint, however, things look different. The European allies see themselves already making commensurate efforts for the defense of the NATO area, whereas they purport to have neither the resources nor the national interests, and certainly not the public support, to be global military powers except for the limited purposes of protecting some independent remnants of their past colonial realms. West Germans, in particular, point to their willingness to draft manpower and to the superior quality of their troops, as well as to their economic aid to Turkey compared to America's relatively poor record of economic aid, as evidence of adequate burden-sharing. Furthermore, the relative proximity of Germany and other European allies to the Soviet Union and their greater dependence on Middle East oil and, hence, on the goodwill of the oil suppliers, not to mention their need to manifest independence from America's global role, militates against their identifying too closely with American policies in the Middle East. Thus they are inclined to argue that the United States exaggerates what military forces in the Persian Gulf can accomplish; that what is needed is more astute and flexible diplomacy, exploiting the special relationships that they have been cultivating with particular countries in the region; and that the United States compounds the problem of fostering regional stability and containment by failing to use its leverage on Israel to secure a settlement the Arabs can accept. On the basis of such competing views of equity there can never be a perfect resolution of the problem of distributing burdens and responsibilities between the United States and its allies.

From an abstract geopolitical and psychopolitical view there is a compelling case for going beyond a reallocation of burdens within the existing structure of military and political responsibilities to effect a radical devolution of power and responsibility for global containment from the United States to its allies. The United States has experienced a steady geographical expansion of its security interests and commitments since the beginning of the cold war but—unlike the USSR, which has more recently bid to catch up with this expansion—no commensurate increase in its capacity to support these interests and commitments, especially where military support is required. Indeed, it operates under what is probably a growing disparity between perceived security interests and real power because of growing constraints, not all of them self-imposed, on its effective power to support containment in the Third World. Its major allies (including Japan) on the other hand have until perhaps very recently experienced a steady contraction of their security interests while the economic base from which they could project power has increased enormously. The United States can no longer hope to narrow the disparity by substituting reliance on nuclear weapons for the projection of local conventional capabilities, building up security surrogates (like Iran), using SALT as a lever to induce Soviet constraint in the Third World, or depending on economic aid to build situations of strength against Soviet penetration. America's allies are the one great remaining asset not fully applied to global containment.

From the standpoint of national psychology, too, the same case for devolution makes sense in the abstract. The argument runs that only fully responsible states will act responsibly. As long as the European allies continue to be so largely security dependents of the United States, they will tend to approach security policies through the prism of how to influence the United States rather than what their own interests require. The corollary of this proposition is that if the allies are released from their status as dependents (by exactly what means, whether the disbandment of NATO or the withdrawal of American troops from Europe, is not clear), they will unite to form their own military coalition, which will then assume a much larger share of the common security burden in Europe and outside it.

National political realities, however, preclude any such dream of devolution under foreseeable conditions. Quite apart from the war-born constitutional/political restrictions that confine Germany (and, more stringently Japan) to a geographically circumscribed military role, the allies (with the partial exception of France) have no incentive to abandon their status as security dependents of the United States and many reasons to retain it. Only perhaps if they were to become convinced that the United States could no longer protect their vital interests, if the United States were determined to help its allies become militarily self-sufficient, and if the Soviets stood idly by, might devolution take place. More likely, U.S. abandonment of its protectorate role would lead the allies to seek security through accommodation with the USSR. In any case, one essential political condition of successful military devolution would be missing: European unity. And one of the greatest obstacles to devolution would remain the fact that in an enhanced European military coalition Germany would be far more powerful in relation to the other European allies than it is as one of the several allies dependent on the United States.

Therefore, whatever readjustment of roles and burdens the allies (essentially to the United Kingdom, France, and Federal Republic of Germany) might undertake in projecting their military and economic power to the Middle East, it will fall far short of relieving the United States of the burdens of discharging the preponderant role of containment. On the other hand a readjustment that completely substituted European for American forces and defense contributions in Europe in order to help the U.S. rapid deployment force would seem calculated to weaken confidence in the American commitment to NATO while impeding any effort to secure a higher level of the allies' cooperation outside NATO. For political as well as material reasons both kinds of readjustment must take place, but they can only be gradual and marginal, within the existing structure of U.S.-allied relations.

Within this structure, however, it is urgently important that the United States strike a new geopolitical bargain: The allies will recognize that the Middle East and Persian Gulf, the NATO area, and also Japan's zone of self-defense constitute a single strategic field, within which they will coordinate the use of their economic, military, and diplomatic assets at a level and in ways commensurate with need and their resources. The United States in return will systematically include the allies in planning and the exchange of information and assessments to an extent commensurate with their assumption of active responsibility. (See the subsequent section on Organization.)

Some Building Blocks

Military Division of Labor. Evidently the United States faces a very difficult and expensive task in constructing in the immediate future the kind of military presence in the Persian Gulf and Indian Ocean regions that can tangibly support the Carter doctrine—that is, an RDF that can really project forces rapidly to the Middle East and fight a major local war, possibly involving Soviet forces—without gutting NATO's (or East Asia's) defense forces. Evidently too the United States can only marginally ease this task by reallocating defense burdens to allies and eliciting their tangible assistance outside the NATO area.

Nevertheless the allies' military cooperation in Europe and the Greater Middle East is indispensable for economic and military reasons and also for political reasons—that is, to demonstrate concerted will to the Russians, Middle Eastern countries, and the American public. In Europe what is needed is for Germany and other allies to pick up some of the slack resulting from the diversion of American efforts to the Persian Gulf by enhancing reinforcement capabilities, reserve forces, ammunition stocks, and logistics support. In the Middle East the United States will profit from British and French cooperation in naval and onshore projection forces, the use of bases and facilities, overflight rights, and transportation. The French have fourteen naval ships operating out of Djibouti and are quite willing and able to use them to protect shipping in the Persian Gulf and even to project small forces rapidly to the land where French interests are directly threatened. The British are contemplating the enhancement of their smaller naval force and the establishment of their own rapid deployment force. The quiet assembly of some sixty U.S., French, British, and Australian warships in the Indian Ocean when the Iran-Iraq war raised the prospect of a closure of the vital Strait of Hormuz suggests the potential for allied operational collaboration. Such collaboration must probably stop short of a joint command, but such a command is not indispensable. Indeed, there may be political advantages to French and British contingents operating independently where they are more acceptable than American forces to countries in the area; for example, the British in Oman.

An agreed division of military labor presupposes agreement on military contingencies and how to respond to them, which, in turn, presupposes a consensus on the kinds of threats to allied vital interests that are posed by the interaction of indigenous and external factors in the greater Middle East. Complete agreement on military and diplomatic strategy is unobtainable, however. What is necessary is sufficient agreement on the general purpose and direction of policies so that differences of national interest do not impede or nullify the expedient actions of the United States or other countries that have the will and capacity to counter threats to the common security.

Diplomacy. Uniformity of allied policies and actions in the Middle East is not possible. But it is not necessarily desirable either, if an ally is better able to pursue a complementary diplomatic line because it is conspicuously acting independently of the United States. Thus during the Iran-Iraq war Britain, France, and West Germany quietly exploited their special diplomatic ties with various Gulf countries in parallel with but in discreet independence from U.S. measures. On the other hand, policies and actions that contradict each other may not only be a diplomatic embarrassment and hindrance but also erode the foundation of broader collaboration.

Differences between U.S. and the allies' policies toward Israel, the Arab countries, and the Palestinians are a case in point. If European deference to Arab oil suppliers' views on Palestinian autonomy would facilitate a stable Arab-Israeli settlement or the establishment and operation of a rapid deployment force, the United States should be grateful for diplomatic assistance from allies that it is constrained from providing itself. Similarly, France's special relationship with Iraq could be helpful insofar as it reduced Iraq's dependence on the USSR. In reality, however, it is hard to see any tangible benefits resulting from allied diplomatic divergencies. European influence on Arab-Israeli relations or on access to facilities is negligible. It may encourage Arab opposition to the Camp David accords and somewhat help the United States to appeal to the Israelis for moderation, but it does not discernibly facilitate a workable alternative. Nevertheless, the United States can hardly afford to turn a deaf ear to its allies' advice on Middle Eastern diplomacy while it is eliciting their cooperation and championing a new era of consultation for the sake of common security goals. U.S. and European differences on the Arab-Israeli issues must not be allowed to foster mutual resentment and parochial propensities that impede the development of a coordinated strategy to support these goals.

Economic Aid and Arms Sales. The division of labor among allies should include a complementarity of allied programs in economic aid and arms sales. The economic aid extended to Turkey by the European Economic Community, following German initiatives, may play a crucial role in rescuing the Turkish economy, stabilizing the Turkish government, and even facilitating a resolution of Greek-Turkish differences in the Aegean and Cyprus. It is politically more difficult and in the conventional view practically impossible to achieve complementary arms sales policies in the Middle East, but systematic consultation on arms sales among the United States, France, and the United Kingdom might lead to weapons specialization and regulation of sales that would advance common political and security goals (for example in Saudi Arabia) without any economic disadvantage, especially considering the desire of a number of recipients not to be entirely dependent on one supplier.

Energy. In the long run the only sure way to minimize the economic and political problems resulting from Western dependence on Middle Eastern oil, including those arising from the different extent of U.S. and European dependence, is for the allies substantially to reduce their dependence through comprehensive conservation and programs of energy alternatives, in cooperation with each other.

Divergences between the United States and its allies with respect to the development of nuclear energy, particularly those arising from U.S. opposition to European programs for uranium enrichment, reprocessing of spent nuclear fuel and fast breeder reactors, would have impeded the process of reducing dependence on oil importation if the allies had not largely evaded or ignored American opposition. The results of the International Fuel Cycle Evaluation (INFCE) confirm the European emphasis on nuclear production rather than the American emphasis on nonproliferation, under what European governments regard as the intrusive terms of the Nuclear Nonproliferation Act of 1978. But the American emphasis persists as an irritant to U.S.-allied relations, even though the United States itself suspended nonproliferation restrictions on the sale of enriched uranium to India for the sake of overriding political considerations. Considering the tangential relationship between European nuclear programs and the prospect or pace of nuclear proliferation, and considering the negligible influence the United States is able to exert on these programs, the United States would be well advised to drop its opposition to European nuclear programs that diminish dependence on Middle East oil and concentrate on positive cooperation, in research and development for example, toward all forms of conservation and energy alternatives.

A better case may be made for U.S. policy on restricting the foreign sale of nuclear reactors and material to Third World states, but again implementation of the policy has been inefficacious or counterproductive. It has tended to sour U.S.-European relations—the futile U.S. opposition to the German sale of nuclear reactor technology to Brazil being the most conspicuous case in point—without significantly affecting the prospect or pace of "nonnuclear" countries developing the ability to "go nuclear."

Regardless of progress in European and U.S. nuclear energy programs, there will remain the short-term task of establishing machinery and procedures for reallocating oil supplies among allies in the event of emergencies arising from the serious blockage of production and transportation of oil from the Middle East. The International Energy Agency (IEA) is supposed to serve this purpose, but few authorities on this subject expect it to work well if it has to be tested. Furthermore, it only comes into effect when there is a 7 percent shortfall in oil supplies.

Some authorities, like Walter Levy, see an even more demanding and equally urgent task of concerting allied oil purchasing policies—for example through an international and national allocation system—in order to prevent very serious economic, financial, and political consequences of competitive panic buying above OPEC prices among the allies in the not unlikely event of substantial reductions of Middle East supplies in relation to demand. But the recommended solution surely strains the bounds of political feasibility.

Economic Sanctions. One of the few nonmilitary levers available to the West to deter, moderate, or penalize adverse actions by the Soviet Union or Middle Eastern countries is economic sanction. The United States has often resorted to sanctions because there seemed to be no other tangible way to punish adverse actions or express outrage. The record of such actions actually affecting the behavior of states is one of almost complete failure, and it is not difficult to see why, given the importance of the stakes to the target state; the inability of the United States to impose serious sanctions because of the unavailability of substitute goods and trade and the difficulty of gaining multinational cooperation, the reciprocal costs to the United States, and the American interest in not completely alienating or provoking the target state. The record of other countries separately or multilaterally using sanctions to change the behavior of offending states is no better. The Organization of Petroleum Exporting Countries (OPEC) is almost a unique exception.

Yet the United States will continue to find the imposition of sanctions attractive, if only for symbolic and domestic purposes. The Iranian imprisonment of American embassy personnel and the invasion of Afghanistan were recent occasions. Events in the Middle East may well provide other occasions. The trouble with symbolic sanctions is that the value of the symbolism quickly depreciates but the costs continue. But they cannot be removed without giving the opposite of the signal intended. Only when removing them becomes a gesture of improved relations may they serve a useful political purpose. Meanwhile the difficulty of inducing allies to cooperate when symbolic distinctions may impose real economic sacrifices on them and needlessly damage relations with the target state only provokes opposition and resentment toward the United States among its allies. This is the case with the post–Afghanistan sanctions the United States has sought from its European allies, even though the economic sacrifice these sanctions would exact was ascertained to be negligible. American pressure on the allies has been almost completely unavailing, but the Soviet Union has exploited the allies' displeasure and irritation with the United States as another lever to divide the United States from its allies.

Nevertheless under some economic conditions economic sanctions enforced multinationally could have a strategic, not just a punitive or symbolic effect. Imposing a sufficient cost on a target state might help deter it from repeating offensive actions. Imposing restrictions on the economy or defense capabilities might not deter but also render the behavior of a state less threatening.

Hypothetically, multinational restrictions on technology essential for the modernization of the Soviet industrial economy, particularly, in the extraction, production, and shipment of oil and gas, might exert a significant moderating effect on Soviet investment in defense or on the kind of Soviet behavior that would intensify the arms competition. The need to acquire such technology from the West, given the inefficiency of the Soviet economy, the Soviet emphasis on economic growth as a measure of national prestige, and high investments in defense, was one of the principal reasons for the Soviet shift to détente. The denial of such technology and the implicit offer to renew it upon good behavior might induce the Soviet Union to revise its priorities in favor of accommodation. The dependence of Soviet growth on energy production, the anticipated loss of Soviet self-sufficiency in oil, the need to develop Siberian gas and oil fields with advanced technology and equipment, and the diminished Soviet economic growth rate may now combine to encourage the Soviets to moderate defense expenditures and even Third World ambitions if the Western and Japanese suppliers of advanced petroleum technology could concert a strategy of export denial. At present, however, nothing short of the Soviet invasion of Poland could lead to such a strategy, and even then impulsive outrage would not necessarily support an effective strategy.

What would an effective strategy of economic denial entail? Policymakers must decide whether the Western interest lies in helping or in hindering Soviet energy production in light of the consequences of Soviet dependence on Middle East oil importation, the effects on the world's total energy supply, the political and economic effects of West German importation of Soviet gas, and other strategic considerations. The links between Soviet energy needs and Soviet defense policies must be analyzed in order to be exploited by the West. To be effective sanctions strategy must elicit the cooperation of the major allies. Consider the problem of German cooperation, for example. A large proportion of the high technology and manufactures the USSR imports comes from West Germany. But although Germany's trade with the USSR amounts to only 2.1 percent of its total exports and an even smaller percentage of its gross national product, a comprehensive embargo on this trade is now politically unacceptable, not only for economic reasons but also because Germany views trade with the East as an integral part of *Ostpolitik*. From the American standpoint the predictable political cost of trying to gain German cooperation in extensive export restrictions seems to far outweigh the problematic political gain from their effect on the Soviet Union.

Under these circumstances the ironic conclusion might be reached that the increase of East–West trade in high technology, although hypothetically a political lever for the West to use against the Soviet Union, comes closer to being the opposite in practice.

In the absence of a strategy to which its allies agree, American economic sanctions against the USSR (or even sanctions imposed collectively by all the allies in an emotional reaction to a Soviet invasion of Poland) can do far more damage to relations between the United States and its allies than the benefits to Western security are likely to warrant. It is a critical area of policy that cries out for objective study and planning among allies before the next crisis preempts a strategic response.

Organization. It is universally agreed among the allied governments that extending NATO's geographical scope to embrace the Middle East or other areas is infeasible and undesirable and that the NATO council is inadequate, although useful, as a forum for discussing world issues. It is also widely agreed that the allies need a structure of consultation and decisionmaking appropriate for dealing collectively with common security problems in the Middle East. If the allies have the wit and the will to come to grips with the foregoing agenda of issues in a spirit of urgent collaboration, they should find no great difficulty in adapting old structures of consultation and cooperation and developing new ones to deal with it.

No single organization will be suitable for all kinds of issues and circumstances or for analytical assessments as well as crisis management and policy coordination. For some purposes, such as coordination of energy policies, a special consultative group including Japan may be needed. Dealing with matters of high diplomacy and military politics in the Middle East calls for different levels of consultation. Annual or biannual summit meetings by heads of state are suitable for the broad assessment of issues. Smaller meetings of key states that are willing and able to assume responsibility are necessary for contingency planning, the continuing assessment of conditions that might lead to crises, and decisions that lead to action.

For broad assessments and general political discussions the principal group of four allies (the United States, the United Kingdom, France, and Germany) plus Canada, Italy, and Japan, which already have an established tradition of meeting on economic and energy questions, constitute the proper format. Other members of the

NATO alliance should be informed and invited to endorse the results of such consultations. For action-oriented meetings of states with active responsibilities in the Persian Gulf area the group of four is the proper body, but it ought to include Japan when matters of economic security and diplomacy are considered. And Italy should join when strategic considerations in the Mediterranean are related to those in the Gulf. The four should meet regularly, more often, and with more staffing and representation of senior officials than the group of seven or other summit conferences. It should report to the governments of other allies. Although its proceedings should be confidential, the existence of its deliberations should be public — even advertised — in order to build domestic support for undertaking the burdens of national efforts. Evidence of the allies collaboration could be particularly salutary in its effect on the American public and Congress.

Liaison with the ten members of the European Community (enlarged to ten by the accession of Greece) should also be maintained. Although their aspiration to play a foreign policy role outside the NATO area, particularly with respect to the Arab–Israeli issues, outruns their ability to achieve any concrete results, and although they lack the underpinning of economic and political solidarity that they would need to assert a coherent "European voice," the ten still represent the best hopes for the kind of broad European coalition the United States has always endorsed. To be sure, in practice Americans view the role of the ten with mixed feelings: it is seen as fine medium of coalescence but something of a nuisance when it opposes U.S. interests and policies. But the advantages of such European cooperation outweigh the disadvantages. The United States made this basic judgment at the beginning of the cold war when in supporting European economic cooperation it subordinated a narrow conception of national interest to the larger benefits of encouraging stronger, more self-reliant and cohesive and therefore more responsible partners. That judgment is even more compelling in this period of growing constraints on American power than it was when America enjoyed economic and military primacy.

The outlook for relations between the United States and its allies that the Reagan administration faces is one of great trouble but considerable opportunity. The current difficulties are serious and deep-rooted. Divisive differences of interest and perception of the multifaceted security threat in the greater Middle East and shape the divergent responses of the United States and its European allies. Yet crisis there is still largely latent. Differences on concrete issues and obstacles to pragmatic cooperation in solving them are not nearly as substantial as the differences of perspective on East–West relations in general. This does not mean that the differences of perspective are unimportant. Rather, they could become the basis for severe antipathies and resentments among the allies, which would jeopardize the security and cohesion of the NATO alliance as never before. But it also means that, given the proper spirit of practical cooperation and compromise, the allies could collectively surmount their general differences, avoid a split, and get on with the hard business of countering the challenging mixture of old and new external threats to their security.

Two realities will fatefully determine which of these two courses the alliance pursues: Soviet behavior and American behavior. NATO has only a limited capacity to affect the first factor; but the new U.S. administration, notwithstanding any decline of American power, can have a decisive effect on the way the alliance responds to the ingredients of the current latent crisis. If it approaches the allies in a mood of frustration and impatience, if it accentuates rhetorical and conceptual differences on détente and containment, applies persistent pressure for compliance against intractable resistance, pursues unilateral courses of action out of exasperation with faintheartedness of its allies, or speaks too stridently to the Soviets while it is building a bigger military stick, the fragile network of allied accommodation could unravel. If, on the other hand it offers the allies a strong and prudent strategic design, consistently executed; if it elicits their collaboration in orchestrating a new transatlantic bargain in which they are genuine partners in the formulation and implementation of this desing; if it elicits cooperation with tact and understanding, while subordinating rhetorical and conceptual differences to the expediencies of coping collectively (though not identically) with specific issues on which real interests converge, it can make the hard geopolitical realities of the 1980s the basis of a new era of cooperation among the NATO allies. This era would amount to an historic adjustment of the alliance to the disturbing developments of the 1970s. The adjustment would be in its way even more impressive than those of the 1950s and 1960s as an unprecedented example of the successful management of a protracted peacetime alliance.

4 THE MILITARY BALANCE
A Staff Study

In Spring 1981 the North Atlantic Treaty Organization marked the thirty-second anniversary of its founding. This period is remarkable as the longest interval of peace among the European nations in this century. Secure in their defense, the nations of Western Europe have been able to devote their attention to the promotion of economic growth and have succeeded in extraordinary measure. As of 1980 the combined gross national product of the ten European Community members of NATO exceeded that of the United States.

The preceding chapters described how this comparative wealth, the natural elevation of each country's efforts toward raising its influence, the ascent of Soviet power and aggressiveness, and differing interpretations of the threat and policies for meeting it have called into question the cohesion and continuity of the NATO alliance. Disagreement among the allies is not a recent phenomenon, however. Other conflicts have occurred, which, given the rather broad margin of strategic superiority enjoyed by the West, could be accommodated without great risk to the alliance.

By any objective measure of the overall force balance, today, however, the military advantage of the Warsaw Pact countries is overwhelming. Cast in NATO terminology, the "triad" of strategic, theater nuclear, and conventional power is no longer superior to that of the Warsaw Pact in any of its dimensions.

The purposes of this chapter are to define just how bad things are, to assess the likely effect of corrective measures already underway (such as the long-term defense program and long-range theater nuclear force modernization program), and unhappily—since all of these put together are not enough to restore a respectable balance—to offer one or two new ideas that may be worth considering.

Action of the magnitude required to restore credibility in NATO as a deterrent to the existing force structure would be unprecedented. A surface hypocrisy exists: The NATO allies want security, know that it requires greater sacrifice but are unwilling to provide the necessary resources. Underlying fear of the Soviet Union is rising in the mind of every head of government in NATO. Communicating that concern to the public and translating it into action is the agenda before NATO. A brief review of how we have reached this point may be useful in dampening undue optimism.

The evolution of United States strategy during the 1950s away from "massive retaliation" and toward "flexible response" is well known. The notion of massive retaliation gradually became subject to the criticism that the threat of a massive nuclear response to a relatively minor political disagreement that did not invoke vital interests of either party was not credible and that the West consequently had to be prepared to deal with smaller conflicts.

The first departure in terms of the posture of the NATO forces involved the stationing in Europe in the mid-1950s of tactical nuclear weapons, originally the Honest John missile and 8-inch howitzer. Improvement of strategy would derive, it was hoped, from the fact that the United States could not attack the Soviet homeland with these weapons and that as a consequence our decision to cross the nuclear threshold would involve less risk and thus be more credible as a policy. The strategic principle was the same, however, as under massive retaliation: assuring our ability to maintain dominance by being able to elevate a conflict to a level at which we would clearly prevail.

The introduction of these weapons into Europe was not entirely welcomed by the allies, whose opposition to the notion of nuclear—or, for that matter, conventional—warfare in Europe was and is unequivocal. Nevertheless the undeniable erosion of the credibility of the former doctrine and the unwillingness of the allies to field a credible conventional deterrent left a void that gave the deployment of tactical nuclear weapons a certain artificial logic. The fact that the role of these weapons and their tactical employment was and is subject to widely divergent interpretations in Europe and the United States tended to undermine that logic, but this deliberate ambiguity was promoted out of political necessity.

During the 1960s NATO's strategy was further refined during the Kennedy and Johnson administrations as additional steps were taken to bolster our capability to deter or fight conventional engagements. The goal, again, was to improve the deterrent value of our military posture, this time by making more believable a U.S. response to any Soviet aggression while lessening the likelihood that any U.S. president would have to resort to the use of nuclear weapons of any type prematurely. It was not planned under this refined policy to match the conventional fighting capability of the Warsaw Pact forces. Rather, the objective was to create sufficient conventional military capability to require a full-scale Warsaw Pact military effort in order to prevail, thus forcing a "pause" that would require the Soviets to contemplate seriously the consequences of continued hostility. Preventing a successful Warsaw Pact blitzkrieg or a situation that would require tactical nuclear weapons to repel a minor military incursion was felt to refine fully, at least in the NATO area, the military policy begun with the theory of massive retaliation. At the conventional level, U.S. forces were to be adequate to deal with modest military incursions and to make obvious the inevitability of a U.S. military response to aggression. Should the USSR nonetheless persist in its aggression, the United States could then escalate the conflict to the level of tactical nuclear warfare and if necessary to the use of strategic nuclear weapons. Since the United States was dominant at both these levels, the Soviet Union was deterred from attack in the NATO area.

Ambiguity clouded certain features of the strategy, but it must be acknowledged that it was successful. Today sober reflection on the reasons for that success lead ultimately to the persuasiveness of the threat represented by the overarching superiority of the United States strategic systems.

Since about 1970, as a result of deliberate decisions, the historic Western superiority at the strategic nuclear level has been allowed to erode toward a condition of parity. Prudence should have dictated that the alliance consider political and military implications of strategic nuclear parity with the Soviet Union, asking what it implied in terms of Soviet behavior and what actions would be required to maintain the balance at the theater nuclear and conventional levels. But no such thoughtful review took place, no new assessment of the threat emerged, and no new allied military programs were charted to maintain lower order deterrence.

To be fair, the United States did not ignore the problem nor the imperative of taking action to assure that the strategic balance did not erode past parity to Western inferiority. The commitments made under the administration of President Gerald Ford to maintaining, if not escalation dominance, at least credible deterrence, were unequivocal. The effectiveness of the American MX missile system to be deployed starting in 1983 seemed for a time adequate to cope with the emergent Soviet intercontinental ballistic missile (ICBM) panoply, which the CIA expected to threaten the U.S. Minuteman force by the mid-1980s. Approval of the B-1 bomber and Trident submarine programs assured the effectiveness of these two legs of the U.S. strategic triad.

At the level of nuclear warfare within the European theater, development of cruise missiles and enhanced radiation warheads promised for a time to preserve the balance. At the conventional level, the Ford administration redirected defense spending priorities in the wake of the Vietnam War to restoring the fabric of U.S. forces committed to NATO. Just doing that—setting aside the need for additional forces—absorbed all available resources.

In 1976, then, there was a reasonable prospect of preserving strategic nuclear parity and some basis for confidence at the theater nuclear level. No serious effort was made within the alliance, however, to strengthen its capacity for conventional warfare although the loss of America's historical dominance in the escalation of nuclear weaponry had heightened the utility of Soviet conventional power and the Warsaw Pact forces were being very substantially improved. As earlier chapters make clear, differentiated détente and economic woes foreclosed any possibility of strengthening the posture of NATO's conventional force.

Since 1976 a number of actions have been taken by NATO and the Warsaw Pact that have affected the military balance at each level. At the strategic nuclear level, better than expected performance and earlier than expected deployment of the fourth generation of Soviet ICBMs advanced the date of Minuteman vulnerability to 1982. The concurrent delay of full deployment of the MX program until 1989 opened the so-called window of vulnerability, almost a decade during which superiority at the strategic nuclear level would give the USSR the escalation dominance the West had held for so long.

At the theater nuclear level, the deployment against Europe of the Soviet SS-20 mobile intermediate-range ballistic missile (IRBM) represented a substantial improvement in the Soviet arsenal. This multiple warhead solid propellant derivative of the SS-16 gave the Soviets a prompt counterforce capability against all of Western Europe, a capability not matched in NATO, thus providing much more discriminating control of theater escalation and by the way rendering all but irrelevant French land-based nuclear systems.

At the conventional level, the Soviet Union had pursued a program of steady improvement of its conventional forces for over a decade. This program had both qualitative and quantitative dimensions although the former emphasis was more dramatic in its effect. During the 1970s, the deployment of the T-72 tank, the ZSU-23 and 57 radar-guided antiaircraft gun system, the BMP/BPR/BRDM family of armored fighting vehicles, the replacement of primitive MIG-17/19 Fitters, SU-19 Fencers, and finally the TU-26M Backfire transformed the Soviet-centered Warsaw Pact troops into a first-class fighting force. Nor was this a qualitative improvement achieved at the expense of quantity; numbers of weapons have increased in virtually every category of weapon system.

In sum the North Atlantic alliance stands today in a state of tenuous parity at the strategic nuclear level with inferiority certain very soon; clear inferiority at the theater nuclear level and lopsided numerical inferiority at the conventional level are also anticipated. The strength of our deterrent and our ability to control escalation by virtue of incrementally superior force are gone, and the situation promises to become worse before it gets better. In offering his assessment of the current balance to the U.S. Senate armed services committee in 1979, the supreme allied commander for Europe, General Bernard W. Rogers of the U.S. Army, had this to say:

> Objective analysis of the growth of Soviet military power confirms its balanced distribution across all major categories of military capability: land, sea, and air; nuclear and conventional; immediate combat power and sustaining logistical support. As a result of this unabated growth of military power, the Soviets have surpassed the West—or soon will—in all three types of forces required by our NATO strategy—conventional, theater nuclear, and central strategic forces.

The reader may protest that quite a bit is being asserted here as fact that is judgment, and a number of things are being done by the allies that hold great promise. This is a fair point, and certainly, such unequivocal judgment must be supported by objective data. Such an analysis follows below. Still, intellectual honesty requires that we approach such an analysis with a clear understanding that recent history provides no basis for optimism. Is allied security better or worse than it was ten years ago? Has the promise in earlier analyses of the combined effects of technological breakthroughs proven real? If not, are we not deceiving ourselves at a time when the loss of nuclear superiority has made time an unaffordable luxury? Though reorganization schemes, added investment, weapon improvements, and even added forces imply that something is being done, historically such measures have often coopted leaders anxious to believe that we will somehow muddle through. It is essential that, after considering the value of the several ongoing programs, we look again at the resultant balance of military power. We cannot afford to be proud of doing a great deal if the outcome remains hopelessly inadequate.

THE CONVENTIONAL BALANCE— LAND FORCES

In the following analysis conventional forces are treated in two categories: those deployed on either side of the border between East Germany and West Germany and thus available for immediate attack or defense and those available for reinforcement of both sides.

On the Warsaw Pact side of the border (East Germany and Czechoslovakia), Soviet forces are deployed as indicated on Figure 4-1. They include in East Germany ten armored divisions and ten motorized rifle divisions, and in Czechoslovakia two armored divisions and three motorized rifle divisions. Added to Soviet forces are seven Czech divisions, of which three are armored, and six East German divisions, of which two are armored. Thus, the total disposable force immediately available in East Germany and Czechoslovakia is thirty-eight divisions, of which seventeen are armored.

All of these divisions are considered category 1 divisions, which means that they possess all of their authorized equipment and between 75 and 100 percent of authorized personnel. Each Soviet-styled tank division comprises 320 tanks. Each motorized rifle divi-

Figure 4-1. Soviet Forces in Central Europe.

sion comprises 265 tanks. Thus within the thirty-eight divisions deployed contiguous to the border between East and West Germany, there are approximately 11,000 tanks. Soviet, East German, and Czech manpower would total approximately 707,000. All Warsaw Pact forces are organized under a centralized, Soviet-dominated command structure; they train together and practice the same military doctrine and tactics and are uniformly equipped with the same weapons.

On the NATO side of the border, Western forces are deployed as depicted in Figure 4-2. The central region is divided into two areas: The northern army group (NORTHAG) includes four corps areas, assigned to the Netherlands, West Germany, the United Kingdom, and Belgium from north to south; the central army group also includes four corps areas, two manned by West German army forces and the other two by the United States.

Proceeding north to south, each corps area is manned during peacetime as follows: The Dutch corps is made up of two divisions, only one brigade of which is actually in West Germany during peacetime. It is estimated that it would require approximately 5 days for the remainder of the two divisions to be deployed, ready to fight, in Germany.

The first West German corps is an extremely professional fighting force comprising four divisions. It is positioned astride the most probable avenue of approach from the east. The British corps consists of four armored divisions, which are manned in peacetime at a strength of 55,000 total troops. In an emergency the British army on the Rhine (BAOR) is to be doubled by reinforcements from the United Kingdom. The Belgian corps consists in peacetime of one armored brigade and two infantry brigades. Thus, a total of eleven divisions comprising 190,000 troops are deployed in the NORTHAG area. Forces assigned to NORTHAG represent approximately one-third of total NATO ground forces assigned to Central Europe.

Remaining forces consist of fifteen divisions—seven German, five American, three French, and a Canadian battle group—which are assigned to the central army group (CENTAG). The French divisions, while included, are not integrated into the NATO military command structure.

Tactical air support is organized to support the land battle in two commands, the second allied tactical air force, which supports NORTHAG, and the fourth allied tactical air force in support of

Figure 4-2. NATO Forces in Central Europe.

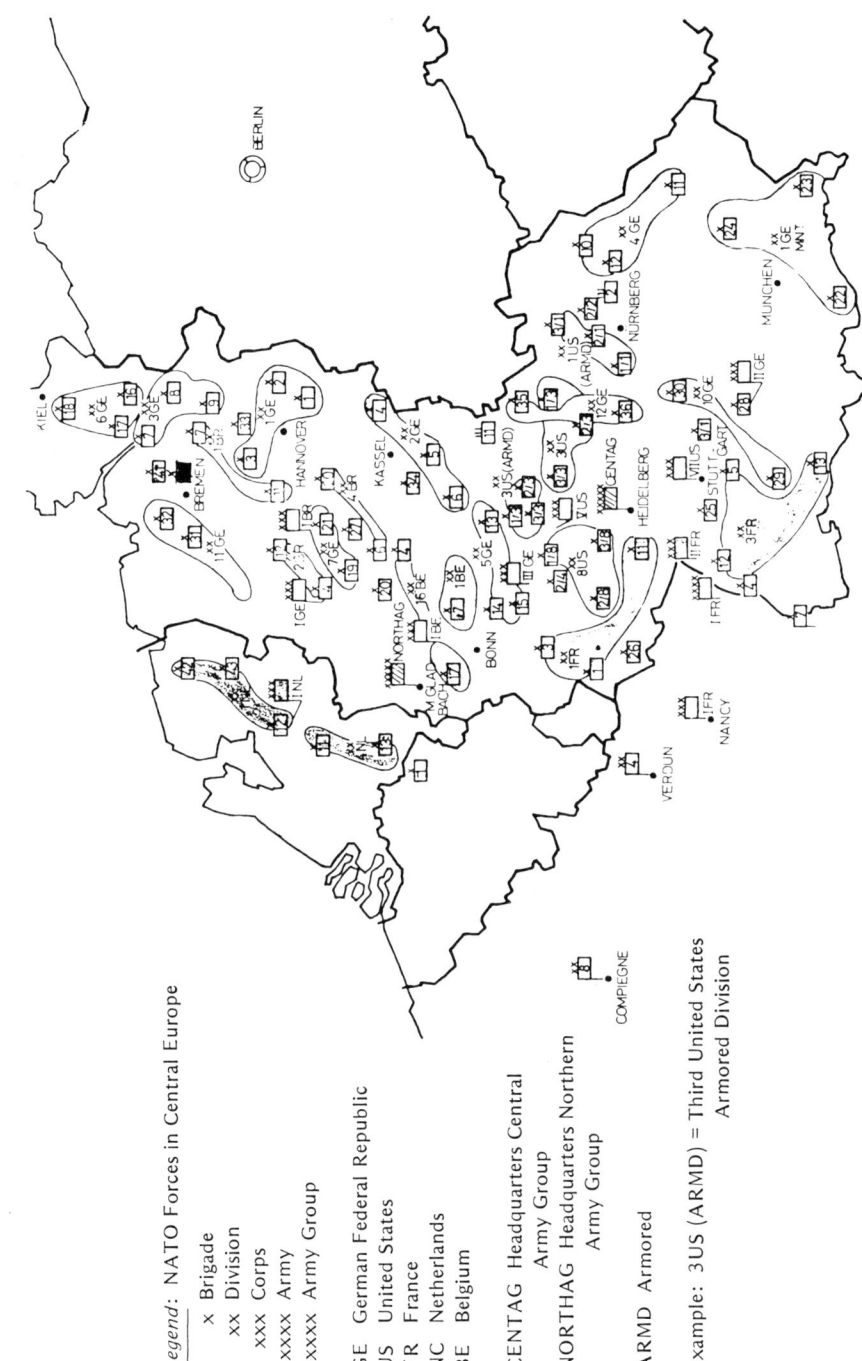

Legend: NATO Forces in Central Europe

  x   Brigade
 xx   Division
xxx   Corps
xxxx   Army
xxxxx   Army Group

GE German Federal Republic
US United States
FR France
NC Netherlands
BE Belgium

CENTAG Headquarters Central Army Group
NORTHAG Headquarters Northern Army Group

ARMD Armored

example: 3US (ARMD) = Third United States Armored Division

CENTAG. Together these two commands comprise approximately 1,000 strike aircraft and 400 others of various categories.

In trying to integrate the doctrinal, geographic, qualitative, quantitative, and other factors that bear on the outcome of general war in Central Europe, one cannot help but be surprised at the outset by several anomalous circumstances that characterize NATO strategy, force structure, and force dispositions. When considering NATO strategy, it is often overlooked that a number of advantages are ceded to the enemy by virtue of the fact that NATO strategy is defensive not offensive. This fact need not be belabored. The initiative granted to the Warsaw Pact has enormous implications, however, particularly when the limited geography available in Central Europe is taken into account. From the NATO perspective, the military requirement is to defend a very narrow strip of land accessible only by sea (West Germany, the Netherlands, and Belgium) against Warsaw Pact forces deployed in depth from East Germany and Czechoslovakia to Siberia.

Geopolitical factors—France's refusal to provide land lines of communication (LOCs) to the West German front and Swiss and Austrian neutrality—require that NATO LOCs run essentially north to south or parallel to the forward edge of the battle area (FEBA). The exposure of these LOCs (5-30 minutes from the front by jet aircraft) plus the suitability of the North German plain for armored operations validates this corridor as a particularly attractive avenue of approach for Warsaw Pact forces. In light of that fact the commander of a homogeneous, well-integrated force would probably choose to deploy his strongest forces and greatest numbers astride that avenue. Setting aside for a moment that NATO forces are not homogeneous or well integrated but, rather, a collection of very dissimilar national forces, it is none the less striking that the forces assigned in NORTHAG to cover this avenue are numerically similar (by a margin of 2 to 1) and less ready than those forces assigned to the more easily defendable CENTAG area. This current disposition of forces is a reflection generally of the final positions occupied by the allies at the end of World War II, but that is an inadequate justification for what is today a fairly striking maldeployment of forces for conventional combat in the FRG. Generally speaking, the European NATO allies have never been particularly concerned about conventional force posture, force disposition, or, except for the Germans, force readiness. This relative indifference is an expression of the emphasis they place on nuclear deterrence as opposed to conventional defense. Today, however, under conditions of strategic parity and a decided Warsaw Pact advantage at the theater nuclear force (TNF) level, to continue to ignore this problem is to send a signal that far transcends military issues.

NATO strategy is based on the concept of forward defense—that is, to contain an attack from the East with the least possible penetration and restore existing boundaries as soon as possible. Traditional military logic, which when considering the threat of a blitzkrieg, would prescribe a strategy of trading space for time must be discarded not only because the space being traded belongs to that member which provides the lion's share of peacetime forces but because there simply is not very much space to trade at all. If the Warsaw Pact forces were to reach Antwerp and Rotterdam, they would have won. Given the criticality of defending the north–south LOCs against a probable attack across the north German plain, it is difficult to understand why the NATO strategy places so little importance on the constitution of an effective strategic reserve within the theater. Warsaw Pact strategy, built around the conduct of highly mobile armored thrusts at points of greatest enemy vulnerability, is well known. Consequently it is understandable as long as our strategy remains defensive that our forces not be so concentrated in one area as to overly expose another. Nevertheless when terrain, trafficability, and the strategic vulnerability of your lines of communication dictate that the most probable avenue of approach will be the north German plain, it makes no sense not only to devote too few forces but to worsen matters by disposing them in completely linear fashion without depth and without a large mobile reserve force. Such a linear disposition without a strategic reserve virtually assures that massed blitzkrieg armored forces would achieve a penetration and exploit it to some depth before forces could be redeployed from CENTAG or from outside Germany to neutralize the penetration.

It is hoped that such a strategic reserve can be created by forces deployed from the United States and in fact a lead brigade from each of three divisions earmarked for NATO will be deployed in the NORTHAG area before 1985. (The first brigade is already there.) It is clear that reliance upon forces that will require some time (approximately ten days to two weeks) to arrive places a premium on receiving warning immediately and acting promptly once warning has been received. To the extent that current NATO and U.S. predictions are that we will have between 8 and 14 days of warning, there is very little reason to be sanguine about the outcome.

This is particularly so in light of the cumbersome nature of the NATO decision process. A fairly comprehensive system exists for monitoring more than 500 indicators of Warsaw Pact mobilization daily and for reporting assessments of these indicators to higher authority. Decisions in NATO are made by a committee, not surprisingly. To expect that the committee could assemble, reach a consensus on the information, and agree to order mobilization in less than 3 days' time is unrealistic.

Once mobilization is ordered a great many terribly complex operations must take place before NATO forces are even disposed to defend. It should be remembered that the ground forces now deployed in Germany are not in their combat defense positions. The majority are more than 100 kilometers from their wartime positions. The time required for the various units to leave their barracks, pick up the necessary ammunition and stores, and move to their field positions varies from 2½ days to a week. It is not anticipated that the Dutch divisions, for example, could move from the Netherlands and be in position ready to fight in less than 5 days.

Besides having to move forces already in Germany, NATO would have to bring forces from outside the country, move them to POMCUS sites to draw equipment, ammunition, and supplies and their final deployment to defensive positions. Amid all of this activity will unfortunately occur such collateral movements as repositioning west of the Rhine units now assigned east of the Rhine, or vice versa. Peacetime political considerations foreclose correcting this malpositioning anomaly at this time.

A final consideration that would affect the ability of the allies to recognize the threat, respond with quick decisions, commence mobilization, and be in position to defend before the attack occurs is the possible disruptive influence of saboteurs and irregulars known to have infiltrated the Federal Republic of Germany in recent years. Some place the number as high as 12,000 to 15,000. The effect of such a cadre in attacking POMCUS sites, ammunition storage areas, and even airfields is incalculable but extremely worrisome.

The foregoing prebattle problems are static. They would be compounded under the dynamic circumstances of battle, for which a number of scenarios exist. The one that follows is derived from the best estimates of informed Western analysts, Congressional studies (principally the Nunn–Bartlett report), and criticism by outside observers such as Belgian General Robert Close.

All twenty of the Soviet divisions in the group of Soviet forces in East Germany (GSFG) are within a 5-hour drive of the border with West Germany, and their readiness is such as to be able to move with sufficient ammunition, supplies, and spare parts to sustain themselves in battle for 5 days within 4 days after being alerted. Prior to alerting their forces to take these more visible actions, the Soviets could have brought these divisions up to full strength in terms of personnel. The GSFG would be augmented by the six East German divisions, which are also category 1 divisions. In Czechoslovakia the total force would consist of five Soviet divisions and seven Czech divisions, all category 1.

Much is often made of the low confidence that should be attributed to East European satellite forces for reasons of loyalty and efficacy. It is the unanimous view of all NATO military staff officers interviewed that Soviet and East German forces are substantially more effective and reliable than other Eastern bloc troops. It is expected that East German forces would fight aggressively in an attack against the Federal Republic of Germany, and it is interesting to note that this view was echoed among leading military officials interviewed in Bonn. Less credence is given to the performance of forces from Czechoslovakia and Poland, particularly in view of internal conditions in the latter country. Nevertheless it is acknowledged that the Soviet forces now deployed in Poland and Czechoslovakia could be replaced by reserve forces from the Soviet Union within 5 days and that consequently the seven Soviet category 1 divisions deployed in Czechoslovakia and Poland would be available for employment in the attack.

In sum, within 4 days of mobilization, the Soviet Union could concentrate twenty-two of its own category 1 divisions (twenty in the GFSG and two from Poland) in an attack across the North German plain. The six East German divisions could be deployed opposite Fulda to conduct operations of sufficient intensity to hold in place German and American forces or to interdict their lateral movement northward. It must be remembered that there is strong incentive for preserving a solid defense in southern Germany during this particular period since the staging areas and POMCUS sites for arriving Reforger and D + 10 units are all in southern Germany. As a consequence no breakthrough in the Fulda corridor can be tolerated. The five Soviet divisions and seven Czech divisions deployed opposite southern Germany could similarly hold in place German forces now deployed in the corps guarding the Austrian-Czech border. The ratio of forces in the NORTHAG area would thus involve twenty-two GSFG divisions against eleven divisions contributed variously by the Netherlands, the United Kingdom, Belgium, and the West Germany. To the extent that NATO forces are deployed in a linear fashion from north to south, the actual force ratio at the point of attack could vary between 3:1 and 5:1 in favor of the Warsaw Pact forces.

THE AIR BATTLE

Approximately 2,800–3,000 Warsaw Pact tactical aircraft would be available to support an air attack. If no problems were encountered in the reinforcement of NATO tactical air force by the United States, the West could count on as many as 1,000 strike aircraft to wage the air battle, seek to establish air superiority, and support the land battle during the early days of the campaign. Thus the Pact would outnumber NATO tactical air forces by a ratio of 3:1. It is undeniably true, however, that NATO aircraft are substantially superior in a qualitative sense. Informed Western analysts believe that the air battle would evolve as follows. For at least the first 3 days of air combat, the Pact would employ its forces in three waves of approximately 1,000 aircraft each, composed and cycled as follows: The first wave would be comprised of 500 fighter bomber aircraft from Eastern bloc countries covered by 500 Soviet fighter interceptors. Their mission would be to attack NATO surface-to-air missile sites, and ground-controlled intercept (GCI) sites. The bloc attack aircraft are limited by range and have no radar guidance capability.

This first wave would be followed within 30 minutes by a second wave, also of 1,000 aircraft, this time all Soviet, divided equally between ground attack and interceptor aircraft. These aircraft would all be modern, high-performance planes, of the Fitter–C, Flogger, and Fencer generation. They would all have a radar guidance capability, and their ordnance could include 500-pound and 1,000-pound bombs as well as electrooptical and TV-guided ordnance. The second wave would be targeted against command and control facilities and shelters. A third wave would comprise 350 Backfire and Badger aircraft, which would be targeted against runways.

It is anticipated that all three waves could be recycled and conduct at least two sorties daily for the first 2 days and that the first two waves could be recycled three times. On the third day the Backfires and Badgers would revert to the control of Soviet long-range aviation (LRA).

Although it is difficult to predict the precise outcome of the battle for air superiority, certain apparent vulnerabilities provide useful data. For example, NATO GCI sites are extremely vulnerable, and it is anticipated that they would be incapacitated the first day, probably by the first wave of aircraft. This would mean that the responsibility for directing NATO intercepts would evolve to the airborne warning and control system (AWACS). The eighteen AWACS aircraft, once deployed, are envisioned to be capable of conducting intercepts simultaneously even when fully jammed.

Informed Western analysts are reasonably confident of the superiority of Western F-15s, F-16s, F-111s, F-4s, and Tornado aircraft over their Warsaw Pact counterparts and, if they are able to overcome the ground-based air defense, will establish NATO air superiority within three to 4 days' time. They are frank, however, in identifying several factors that could alter not only the time required but the overall outcome.

Most important is the very substantial threat posed by Soviet ground-based air defense systems, the foremost being the improved SA-6, the SA-8, and the radar-controlled ZSU-23 antiaircraft gun. Effective neutralization of surface-to-air missiles requires electronic jamming and attack. Based upon the organization of SA-6 units (one radar per six launchers), there has been a requirement for one wing of F-4 (Wild Weasel) electronic warfare aircraft. The SA-8 features a radar integral to every launcher. As a consequence, the requirement for electronic warfare aircraft on the NATO side is increased by a factor of 5.

It is the judgment of senior air force planners that under current conditions NATO can be reasonably sure of being able to blunt a Warsaw Pact air attack. They are confident that after three days the Pact will have lost the capability for a massive air offensive although they will still be able to launch a substantially reduced strike on a given day. This outcome is premised on the availability in NATO of six wings of F-15s and six wings of F-16s.

Four additional factors could alter this assessment. The first is the well-known problem of weather. The prevalence of poor flying weather (instrument metereological conditions) over Germany is well-known. Within the NATO arsenal only the F-111 approaches having a reliable all-weather, night-attack capability. Efforts to improve the Western capability are proceeding but with no real basis for optimism in the near future. In order to live with the problem, procedures have been developed for more flexible allocation of sorties throughout the theater from whatever launchpoint may be available and operable on a given day; that is, that although the probabilities of visual conditions existing throughout Germany on a given day in January may be below 10 percent, the probabilities on the same day that at least one airfield will experience visual conditions may be over 90 percent. As a consequence, it would be worthwhile to have the flexibility to be able to assign aircraft from any operating airfield to the sector where they are most needed on a given day. At best, however, this would translate to a very substantially degraded air combat and ground support capability more than 60 percent of the year in Germany.

A second factor that must be taken into account is that the aforementioned outcome is conditioned on the employment of virtually the total NATO tactical aircraft inventory. The uncertain stand-off achieved would be reversed immediately if the Warsaw Pact forces were replaced, for example through the redeployment of substantial tactical air force assets now deployed in the Far East. As was mentioned in earlier chapters, a great deal of the speculation as to the outcome of a NATO/Warsaw Pact conflict is conditioned upon assumptions of various levels of Chinese activity. This has particular relevance in the air battle. In short, if the tactical air assets now deployed opposite China were moved to the European theater, the Warsaw Pact could establish and maintain air superiority over the West.

A third factor that could affect the outcome is the possible effect of preliminary attacks against NATO airfields by saboteurs and infiltrators. This matter receives very little attention in NATO but represents a very clear vulnerability. Moreover, inadequate security is an even sharper problem with respect to nuclear storage sites in Europe. Quite apart from the implications for general war, the vulnerability of these sites to terrorist attack today has reached very dangerous proportions and should be corrected immediately.

A final factor that could significantly influence the outcome of the air battle concerns the Warsaw Pact's possible use of chemical agents against NATO forces. Because of the inadequate state of unit readiness and training for defense against these attacks, there would be a substantial effect on sortie rates. In the judgment of well-informed observers, persistent use of chemical weapons by the Warsaw Pact would virtually assure that NATO would lose the air battle.

Turning to the significance of the battle for air superiority on the ground battle, it is first apparent that for at least two to three days, air support of ground forces will be almost zero. As Professor Luttwak has pointed out, In the first few days of a NATO war when airpower would be needed most to give time for the ground forces to deploy, the U.S. Air Force would, in fact, be busy protecting its own ability to operate at all. Although most knowledgeable pilots and analysts give the A-10 high marks as a close support aircraft, all acknowledge its vulnerability to Soviet air defenses. The twenty-two divisions postulated for the Soviet attack across NORTHAG would contain hundreds of surface-to-air missiles. Until these are substantially eliminated, the A-10 could not operate nor do most analysts give NATO helicopters much likelihood of survival against the main attack force in the early days of combat.

If all goes well and air superiority is established, thus enabling a diversion of air assets to support of the land battle, what effect would there be? Remember, now, that this is D + 4 and the Dutch, British, Belgian, and German forces are deployed linearly from north to south trying to contain the Soviet attack. Again, the Soviet attack could draw upon as many as twenty-two divisions; but for reasons of pure congestion, these forces would likely be split into at least two attack forces. Assuming that either of these forces chose to attack in the best-defended area (the Magdeburg–Hannover–Osnabruck axis defended by the British and German corps), they would still enjoy superiorities of the following magnitude: manpower, 1.5 to 1; tanks, 3 to 1; armored fighting vehicles, 2 to 1; artillery, 4 to 1; and mobile air defense weapons, 6 to 1. When antitank guided munitions available to both forces in this particular area are added, the tank/antitank balance between the two sides improves to approximately 2 to 1, still substantially in favor of the Warsaw Pact. Thus the force ratios at the outset of the battle are far from encouraging. This fact of life, long acknowledged by NATO, has been rationalized by our ability to reinforce or to escalate the battle through the use of tactical nuclear weapons. Problems involved in reinforcement will be explored more fully in the next chapter. Suffice it to say here that our ability to reinforce, given anticipated SLOC and APOE interdiction, under the best of circumstances does not match that of the

Soviet Union. In general terms we hope that if all goes well, NATO will be able to receive five U.S. divisions within 10 days. That is a goal NATO hopes to meet by 1985, but the Soviet Union has the capability today to augment the twenty-two-division attack force with an additional twenty-nine divisions from the Western military districts of the Soviet Union and approximately 500 additional tactical aircraft by D + 10. On that day, our incremental addition of five divisions would have been offset by a corresponding addition of forty divisions by the Soviet Union.

That a certain amount of calamity seems inevitable for any NATO mobilization but not Warsaw Pact action is surely unreasonable. The high annual turnover of two-year conscripts in the GSFG is bound to have an impact upon force readiness; nor is it certain that GSFG troops could react to an alert with the speed implied earlier. In addition, little things like traffic control and the coordinated phasing of convoys could represent substantial problems. The evidence provided by the Soviet invasion of Czechoslovakia in 1968 was not impressive. Furthermore it should be stressed that Warsaw Pact forces never exercise or practice movements on the scale discussed in this chapter.

Unfortunately many of the same things could be said about NATO forces, which suffer the additional problem of command and control. There are obvious difficulties in coordinating and commanding under combat conditions an army that speaks six languages, does not employ common weapons systems or communications, does not have colocated air and ground headquarters, has very little secure (nonjammable and interceptible) communications, has seldom exercised for extended periods as a fighting force, and lacks the manning for 24-hour combat operations. This very poor state of allied command and control must be compared to the unity of command, standardized equipment, and common doctrine and tactics enjoyed by the Warsaw Pact.

RESTORING THE BALANCE

This recital of the several problems that compromise the credibility of deterrence and defense at any level by the North Atlantic Treaty Organization is not new. Those associated with maldeployment, malpositioning, command and control, and standardization of equipment date from the birth of the alliance. During the 31 years of NATO's existence, adequte improvements have not been made.

To be fair, however, the awareness of the shift that has taken place in the strategic and theater nuclear balance and the manifest importance associated with that shift for strengthened deterrence at lower levels have resulted in a greater level of interest among the allies in corrective action. Two programs have been adopted that should result in substantial improvement in a number of areas. These are the long-term defense program (LTDP) adopted at the May 1978 NATO summit, and the program adopted in December 1979 for the improvement of NATO long-range theater nuclear forces (LRTNF). The LTDP concept was proposed by President Carter at the May 1977 NATO summit. It was reviewed by defense ministers and approved the following May. It includes measures designed to improve the strength of the conventional deterrent over the next 10 to 15 years. The several hundred proposed measures are broken down into nine functional areas. Five are associated with improving capabilities of forces already in the field (readiness, air defense, naval posture, command, control, and communications, and electronic warfare). Three deal with measures to achieve a faster buildup of strength during times of tension and to sustain combat (reinforcement, reserve mobilization, and logistics). The ninth measure deals with how to better allocate (rationalize) available resources among the allies.

The readiness program addresses itself to improvements in the NATO alert system, to improving armor and antitank capabilities, and to improving NATO readiness to deal with chemical and biological warfare. The air defense program focuses on increasing the inventory of missiles as well as the number of personnel devoted to maintenance. The naval program focuses on improving the ability of the individual naval components to operate together. Better communications, better fleet air defense, better antisubmarine warfare equipment and procedures, and improved mine warfare capability are among the basic objectives.

The command, control, and communications program (C^3) is fairly self-explanatory. It seeks to establish a greater degree of commonality in doctrine, procedures, and organizational structure within NATO. It is clear that much more could be done in this area, however. The headquarters of allied forces for Central Europe, for example, has virtually no control during peacetime over the activities of the several member nations' forces. Being under its own national command, each of these seldom engages in large-scale joint training, nor does the headquarters have the charter to bring together the several component nation contingents into a truly combined planning system. It is in short a collection of individuals with a great deal to do, unable to do very much and representing an enormous vulnerability in the event of combat.

The electronic warfare (EW) program deals with ways in which NATO can improve its capabilities for dealing with an ever-improving Soviet EW potential across the spectrum of active and passive measures. This is an area of particular NATO vulnerability. Because of the fairly substantial costs involved, enthusiasm among member nations is quite low. As a consequence the state of preparedness for electronic warfare is extremely low throughout NATO and is not expected to improve substantially as a consequence of the LTDP.

The reinforcement, reserve mobilization, logistics, and rationalization programs will be dealt with in greater detail in succeeding chapters. The interest reflected in the creation of the rationalization program is particularly welcome. It reflects the fairly common undercurrent of opinion within NATO circles that "you can't get there from here" with respect to the gross imbalance existing in conventional power. One cause of this condition is the expenditure of enormous sums for what is often redundant research, development, and production of similar weapons systems. No one ignores the political and economic implications of trying to breach this condition that is inherent in any organization composed of separate nation-states, nor does anyone expect that we will achieve "perfect competition" or a remotely equitable division of labor in the near term. At least there has been a recognition that when the alliance is spending more than its adversary on defense and is not keeping pace, it needs to make better use of its resources. Whether or not this recognition will be translated into improved cooperation in the production of defense systems remains a critical question.

Almost three years after the adoption of the long-term defense program, it is reasonable to stand aside and make an objective judgment as to whether it is working. The first status report on each country's progress in meeting the LTDP goals was submitted in July 1979. Very spotty but nonetheless positive progress was made by each member country. A more telling demonstration of the degree to which the program is regarded as extraordinary and deserving of special emphasis is reflected in the fact that by the end of 1979 all countries had integrated the LTDP improvement measures into their normal annual defense planning cycle, and although LTDP measures generally took precedence, they were not considered sufficient to justify additional funding. That is, for every LTDP measure agreed to, some other previously planned action was dropped. Stated another way, each member country accepts in principle the NATO-adopted goal of increasing its defense budget by 3 percent in real terms each year. It is clear, however, that any extraordinary measures subsequently agreed to, such as the LTDP, will have to be accommodated within that overall ceiling. Consequently the LTDP represents a shifting of priorities in the allocation of defense funding but no net increase in defense investment.

Following the Soviet Union's invasion of Afghanistan, a proposal was introduced to accelerate certain measures within the LTDP so as to achieve earlier improvement in several categories of readiness. At the Spring meeting of the NATO ministers in May 1980, a two-phased program was adopted for pursuing these accelerated actions. To be fair, these represented more than cosmetic measures necessary to mollify the United States who had taken the lead in proposing vigorous action to respond to the USSR's more aggressive behavior. The measures in question were selected by the major NATO land and sea commanders. Following review by the NATO military committee, the measures were referred to each country's government as a basis for eliciting national approval preliminary to the adoption of the accelerated programs by the NATO ministers in December 1980. As mentioned earlier, the LTDP had become a fairly routine matter, thus it may be said that the actions taken in May and December 1980 restored a certain sense of urgency to the need for strengthening the deterrent. Nevertheless it is already clear from the tentative views reported to the NATO headquarters by each nation's government that these accelerated measures will also have to be accommodated within the overall ceiling of 3 percent annual real growth, and even there, it has become clear as a separate matter that many nations will not be able to reach that goal.

The NATO military staff also noted that there are a number of very serious shortcomings in Western defense that will not be corrected by the LTDP or the accelerated program. For example, in electronic warfare the LTDP measures do not go nearly far enough. The allies' responses are inadequate because of the costs involved and the unavailability of sufficient trained personnel. The staff also noted that the national replies to the two-phased program of accelerated measures represented "agreements to consider" the actions in question, not commitments. Again, this was due to national uncertainty as to the availability of the financial resources to pay for the required actions.

With respect to the improvement of reserve forces, the measures contained in the LTDP are very long-term measures and will only result in the addition of one reserve brigade to the NATO force structure. The NATO military staff laid particular stress on the fact that, even assuming complete fulfillment of the LTDP and the accelerated programs, the collective effect will not be enough to restore a credible conventional deterrent or warfare capability.

Within the United States, the LTDP is being taken seriously, and a number of very constructive programs are underway. Included are such measures as increasing the number of weapons in artillery batteries by 33 percent, increasing procurement rates for antitank guided munitions (TOW and Dragon). By the end of the Carter administration, U.S. forces in Europe had over 50,000 TOW missiles. Other measures include the development of a general support rocket system to provide additional area firepower to U.S. forces, continued development of the Copperhead laser-guided antitank artillery projectile, increasing the level of war-reserve artillery ammunition in Europe, and increasing the manning of divisions based in the Continental United States (CONUS) but earmarked for Europe so as to avoid delays that would be necessary to fill out understrength units. Great voids remain, however, even in the U.S. program. These include a woefully inadequate electronic warfare program and hopelessly inadequate training and procurement to cope with combat in a chemical or biological environment.

The Role of Precision-Guided Munitions (PGM). As interest in the conventional force balance rose during the 1970s, one of the frequently cited strategies for redressing it without the burden of placing additional troops in the field was to apply superior Western technological know-how to the problem. Enthusiasm for this approach to the defense against tanks has attenuated since the 1973 Arab–Israeli War, where PGMs were first used in fairly high-intensity armored warfare. Reduced to its essentials, the hope deriving from that conflict was that the greater firepower of PGMs when deployed in a defensive mode would blunt the high-speed armored offensive by requiring the opponent to separate its massed forces lest they be destroyed in the face of a now-lethal defense.

Careful analysis of the Arab-Israeli War and more extensive testing, which has identified more clearly the limitations of PGMs, leads to a less optimistic estimate of their real value. A number of problems exist with the weapons themselves. Most require that the operator maintain line of sight with the target during the entire flight time. Terrain and weather often make such contact impossible. Moreover countermeasures such as smoke, jamming, and evasive maneuvering also degrade the performance of PGMs, and the Israelis have also demonstrated that the effects of direct hits by PGMs can be minimized by structural and design changes in the tank itself.

Other limitations of the current generation of PGMs include their relatively slow rate of fire and vulnerability to enemy suppression fire. Ground-based PGMs are currently delivered from exposed platforms that are extremely valuable targets and that are likely to receive priority of fires from artillery as well as direct fire weapons. Much of the advocacy of PGMs is derived from the experience of the 190th Israeli armored brigade, which was virtually destroyed on October 9, 1973, in large measure because of PGMs. The conclusion with respect to the "revolutionary value" of PGMs appears to have been overdrawn, however. The real instruction of the Middle East War of October, 1973, is not to be drawn from the early slaughter in the Sinai of a single, unsupported, and incautious Israeli tank brigade. The true lessons may be extracted from the knowledge that, of the approximately 3,000 Arab and Israeli tanks destroyed or damaged during the war, at least 80 percent were knocked out by other tanks. The use of hundreds if not thousands of anti-tank guided missiles by both sides exerted at best a marginal influence on the outcome of the ground battle. Another criticism of overreliance on PGMs concerns their susceptibility to instant obsolescence. Weapons that rely upon sophisticated radio telemetry are vulnerable to jamming and other countermeasures. Ignorance of such measures, accompanied by continued confidence in the efficacy of such systems, can have an extremely unfortunate outcome at the moment of truth. The SAM-2, which was responsible for downing the American U-2, was neutralized during the Vietnam War. Similarly, the STYX naval missile, which astonished sailors in 1967 when it sank the Israeli destroyer Elath, was completely ineffective by 1973. In the opening days of the October war, the Egyptians fired more than fifty STYX missiles at various Israeli patrol boats. Not one hit was made. Meanwhile the Israelis were able to sink a score of Syrian and Egyptian vessels and establish control of the sea. Given the size and sophistication of electronic research and development now in progress in both the Soviet Union and the West, the effective life of a new technology promises to become shorter and shorter.

What can be concluded from the discussion thus far? Generally speaking it seems fair to say that the evolution of the military balance toward a condition of parity at the strategic nuclear level and Warsaw Pact superiority at lower levels requires a determined NATO effort to restore the strength of its deterrent at both the theater nuclear and conventional levels. As for the effect of those efforts at the conventional level, however, it appears that, although they have had a constructive effect in terms of forging improved staff cohesion, their effect in fundamentally altering the conventional military balance is marginal at best. Before making final judgments, however, it is necessary to examine measures underway to restore the TNF balance in view of the key role these systems have played historically in preserving NATO's ability to control escalation.

THEATER NUCLEAR BALANCE

Today, the NATO theater nuclear force posture consists basically of approximately 7,000 nuclear warheads that can be fired from about 1,000 ground launchers. These systems are essentially low-yield, short-range weapons whose effect is limited to the European battlefield. The Pershing I missile, with a range of approximately 400 miles, is the most capable system. These ground-launched systems are augmented by approximately 400 nuclear-capable F-4 aircraft as well as carrier-based aircraft deployed in the Mediterranean. Also positioned in Europe are 150 F-111s and approximately 400 Poseidon warheads assigned to SACEUR for use in the general defense plan. Finally, the United Kingdom contributes fifty-six Vulcan bombers and four Polaris submarines with sixty-four nuclear missiles.

The Warsaw Pact TNF force consists of some 1,500 battlefield weapons with ranges up to 500 miles and some 600 SS-4 and SS-5 IRBMs with a range in excess of 2,000 miles targeted against Europe. As with NATO, these systems are augmented by approximately 750 bombers and scores of nuclear-armed tactical aircraft with ranges of up to 500 miles.

This roughly equivalent balance was altered significantly in 1977 by the Soviet deployment of the SS-20 IRBM and the Backfire bomber. The SS-20 is a highly accurate, mobile, MIRVed system that carries three warheads capable of attacking military and civilian targets throughout Europe from deployment areas in the Soviet Union. It is relatively invulnerable to counterattacks by NATO forces.

The requirement to respond to this significant alteration of the balance was first raised by Chancellor Helmut Schmidt in October 1977:

> SALT... neutralizes [Soviet and U.S.] strategic nuclear capabilities. In Europe, this magnifies the significance of the disparities between East and West in nuclear, tactical, and conventional weapons.... We must maintain the balance of the full range of deterrence strategy. The alliance must, therefore, be ready to make available the means to support its present strategy... and to prevent any development that could undermine the basis for this strategy.

Chancellor Schmidt's concern has been translated into a specific program for long-range theater nuclear force modernization (LRTNF) built around the deployment of 108 Pershing II missiles and 464 Tomahawk ground-launched cruise missiles.

The Pershing II is a supersonic, highly accurate mobile missile with a range of over 1,000 miles capable, when deployed in the FRG, of striking targets in the Western part of the Soviet Union, as far east as Kiev. The Tomahawk is a highly accurate subsonic missile with a range of about 1,500 miles. As agreed at the NATO ministerial meeting in December 1979, the deployment of these systems will commence in 1983 and will be completed by 1988.

Without rehearsing the reasons for support for and opposition to the LRTNF program among our allies, now that the decision has been made, it is appropriate to ask how it will affect the balance. Currently the USSR has approximately 900 weapons deployed within striking range of Western Europe while NATO has only about 226 systems capable of reaching the Soviet Union. This provides the Warsaw Pact with a roughly 4:1 advantage over NATO in warheads and a 3:1 lead in equivalent megatonnage. It is anticipated that by the mid-1980s, the Soviets could increase the number of delivery vehicles to 1,300 by additional deployments of SS-20s and Backfires.

Against that number, even with the deployment of 572 Pershing IIs and GLCMs, the comparative Soviet advantage in warheads and equivalent megatonnage will increase and the Warsaw Pact will achieve a 2:1 advantage against hard targets. Beyond 1985, even assuming completion of the NATO LRTNF program, the Soviet advantage will grow even more. It is possible that the Soviet Union might conclude that, if they were to escalate to the theater nuclear level, the West would be unable to respond with adequate force and would be faced with the choice of either unleashing its strategic nuclear forces or conceding. Such a perception could have a very destabilizing effect.

This perception and the rationale generally for the LRTNF program is not uniformly shared throughout NATO. Consequently the decision to deploy even this limited number was probably the most that could be expected at this time, and it was conditioned upon the requirement that efforts be undertaken to engage the Soviet Union in arms control negotiations directed at reducing theater nuclear weapons on both sides. To the extent that such negotiations would take place in an environment in which the Warsaw Pact already holds the advantage in terms of deployed systems, and even the full implementation of LRTNF will leave the West at a disadvantage, NATO's leverage is considerably undermined.

A separate effect of the Soviet deployment of the SS-20 and the Backfire concerns the future of the French nuclear force, the vulnerability of French land-based, missiles and the entire credibility of the French nuclear force posture as a deterrent. The French government has made a decision to improve its submarine launched ballistic missile (SLBM) capability, but this will not affect the fundamental credibility of their strategy now that their forces have been reduced to a modest, assured destruction capability. It may be that this offers an opportunity for better integration of France into NATO military planning.

ALTERNATIVES FOR THE FUTURE

In light of the foregoing pessimistic evaluation of trends in the military balance at each level, the question becomes what to do. At least three possibilities are apparent. One is to try to turn back the clock—to restore the strategic nuclear balance to the relationship that existed throughout the 1950s and 1960s when the United States maintained credible strategic nuclear superiority. In view of the apparently inexorable pace of Soviet strategic programs and the very substantial political and economic impediments the United States would face in such an undertaking, the feasibility of this option is extremely uncertain, but on the whole, unlikely. Even if feasible, the length of the development cycle makes it at best a long-term solution.

A second approach would involve a fundamental reorientation of NATO military strategy. Currently NATO military strategy and force posture are devoted to the conduct of a successful forward defense, or if that fails, to preserving the demonstrable ability to control escalation through the theater nuclear and strategic nuclear levels of the force spectrum. In view of the erosion that has taken place in the credibility of this strategy at all levels, there may be value in considering the wisdom of departing from this essentially attrition-oriented approach toward what some have called a strategy of maneuver. Reduced to its essentials, such a strategy would have both a theater and global dimension. At the theater level, it would probably not be possible to adopt a doctrine of classic maneuver. Not only is the concept of trading space for time politically unacceptable but for as long as France remains aloof from NATO military planning it would be geographically infeasible. NATO is politically constrained from maneuvering backward and politically and materially constrained from maneuvering forward. Nevertheless a number of tactical and structural changes could be made in the NATO force posture to enhance its ability to contain and repel a Warsaw Pact attack. These involve resolution of problems associated with command and control, maldeployment, malpositioning, and the absence of an effective mobile reserve force.

In its global dimension, the new strategy would focus upon the need to make clear that an attack against the West would not be countered by forward defense in Europe alone, but as well by military attacks against targets valued highly by the Soviets and points of vulnerability in the Soviet homeland.

The difficulties to be overcome in promoting and ultimately adopting such a dramatic alteration of strategy cannot be overstated. In reflecting on the very complex and lengthy deliberations associated with past changes in NATO strategy at a time when these debates could be safely conducted beneath a superior nuclear umbrella, one can certainly not be optimistic about the far more tortuous effort that would be required to effect such a change today.

A final approach might be called "improved muddling through." This approach would proceed from the premise that NATO possesses financial resources superior to those of the Warsaw Pact and that it has both the will and superior means to defend itself. It is based on the thesis that it can achieve credible deterrence at every level by making more efficient use of the resources at its disposal. Some would say that this is the approach that has been adopted and that is in full swing, as represented by the LRTNF and the LTDP programs. The point of this discussion, however, is to make clear that such efforts will be fruitless, resulting only in self-delusion accompanied by further erosion of the military balance. This is not to say that these programs are not worthwhile. In addition to LRTNF and LTDP, the efforts being devoted to the development of an effective allied process for the definition of weapons requirements, the division of labor for research, development, and production of new weapons systems, and the elimination of redundancy in defense expenditures are enormously worthwhile. We simply must achieve a greater collective output from the input of individual nations.

An unstated but underlying assumption of this discussion has been the ability within NATO to forge a common perception of the threat. For reasons elaborated in Chapters 1 and 2, it is not at all clear that this premise is valid. This chapter has dealt with Warsaw Pact *capabilities*; differing interpretations of Soviet *intentions* would, of course, alter the conclusions.

5 PROBLEMS OF READINESS, REINFORCEMENT, AND RESUPPLY

George C. Blanchard,
Isaac C. Kidd, Jr., and
John W. Vogt

There is an old tale of two leaders who had to lead separate parties across a wide stretch of bad lands. At the end of the first day, one wrote in his diary: "Twelve hours of marching, only half way into the bad lands." The other's entry was "already halfway out of the bad lands after only 12 hours' march."

The litany of deficiencies recited in this chapter should by no means lead to despair at the size of the tasks still ahead. Rather, it should be viewed as pointing the way to a future that is by no means beyond the possible. Nonetheless, the gap between Warsaw Pact and NATO strength is widening dangerously.

NATO has come a long way since the end of World War II. A major power no Soviet planner can take lightly, NATO is already more than halfway out of the bad lands. Facing up to what remains to be done is the first step in going the rest of the way.

Readiness, reinforcement, and resupply determine the ability of the NATO alliance to implement its war plans. In this chapter discussion focuses upon the NATO central region, but that does not mean that the northern and southern regions are unimportant. Indeed, they may be more vulnerable, an initial attack may be more probable in either, and the problems of war plan implementation may be more severe. However, the problems of the center are indicative of those of the entire alliance and are generally applicable to all three regions.

It is in the center where the Warsaw Pact has deployed forces near the interzonal border far stronger than required for defense. Their force ratios range from 2:1 to 4:1 or even 5:1 compared with NATO, in mechanized armies that can, with air support, rapidly advance toward the English Channel ports unless NATO's armed forces are ready and can be reinforced and resupplied rapidly.

This examination looks principally at NATO's armies and air forces, recognizing the importance of navies in reinforcement and resupply of the land and air battle. Force capabilities of the navies of NATO are reflected generally in the discussion of armies and air forces. Keeping the sea lanes open is essential, and supplies, equipment, and manpower must cross the Atlantic for support of any conflict lasting more than a week or two.

A simultaneous crisis in another area such as the Middle East must also be considered. In such a so-called 1½ war scenario the United States must be able to project and support combat forces concurrently in two widely separated geographical areas. Planning combat in such a situation would be difficult because, for example, some divisions and air squadrons would be on the troop lists for both contingencies. Even more difficult is the fact that many of NATO's service support units also appear on *both* troop lists, and little flexibility exists in their deployment.

Common sense might suggest that the solution lies in the United States' requiring its NATO allies to "do more" in Europe if it is to fight the non–Communist world's battle elsewhere, say, the Middle East. Although this rationale has merit and discussions are proceeding in this direction, limitations exist that are difficult to overcome on both sides of the Atlantic. Primary problems involve the long distances from the United States, competition for available air and naval lift, scarcity of few-of-a-kind logistic support units and the types of "do more" that are feasible for our European allies.

READINESS

Readiness depends upon resources of trained personnel, modern operable equipment, available reserves of both, and operational interoperability with both U.S. and allies' reserve forces. For some time NATO in-place general purpose forces, though high on the U.S. Department of Defense (DOD) priority list, have been living with an inadequate allocation of scarce resources in personnal and dollars. Fortunately the Reagan administration intends to increase the dollars. Unfortunately an extended recovery period will be needed before past deficiencies are made up.

Readiness of U.S. Forces Stationed in the Continental United States (CONUS)

Much has been published describing the viability or lack thereof of NATO's armed forces personnel and in particular the U.S. volunteer forces in both numbers and quality. With the exception of the United States, the United Kingdom, and Canada, NATO nations use conscription to keep their units up to strength and to provide trained reservists. Draft terms in these countries run from one to two years and have in general resulted in effective forces. In the United States, the air force and navy have been able to maintain their numbers consistently; the army, only sporadically. As of 1981 all services were recruiting close to established goals.

Problems exist in all military services, however, in retaining qualified career leaders and specialists, too many of whom leave the service to accept more attractive offers in the civilian economy. The problem has been compounded by pay scales that are lower than those on the outside, extended tours of duty away from families, particularly overseas, and a feeling on the part of many that the United States does not support its armed services well. Recent pay hikes, promises of more, and indications of a return to some type of GI bill of rights, all will undoubtedly be helpful over time.

The U.S. army, for example, has reached its recruiting goal in total numbers of personnel with a higher proportion of high school graduates. Hidden within the happy 100 percent of total numbers, however, is the reality of maldistribution; there are too few noncommissioned officers and specialists and too many privates and PFCs. And the situation has not clarified despite the pay raise of 11.7 percent that took effect October 1, 1980.

The U.S. air force generally has met enlistment requirements. It has had more of a problem with reenlistment: Trained specialists are tempted to leave for more attractive offers outside in defense industries or airlines, which has meant costly training of new personnel. In effect, the air force and navy have been serving as a training base for the private sector. One effect of this phenomenon has been a reduction in experience levels among services personnel with a concurrent impact on combat readiness. Morale has also suffered as units struggle to achieve prescribed readiness levels. As is the case with the army, recent increases in pay and benefits have moved to reverse the trend. Recent slowdowns in airline operations have also been a factor in reducing the pilot retention problem, which for the last several years has been a severe one.

Table 5-1 portrays the shortages of noncommissioned officers and petty officers in the services. A greater incentive must be provided to American youth so that a wider and more equitable representation of American society is attracted and retained in the military. Increased problems in the not so distant future can be foreseen as the impact of lower birthrates makes itself felt in the number of available 18-year-olds. In 1986 the armed forces will need to attract one out of every four or five physically and mentally qualified high school graduates available if present armed forces strengths are to be maintained.

Officer readiness and retention is problematical because too many officers resign in midcareer, particularly pilots in the navy and air foce. Family pressures mentioned before, resulting from separations due to overseas tours, as well as higher pay and career opportunities in industry particularly for specialists, cause too many highly qualified officers to depart. It is hoped that pay scales more comparable with the civilian workforce and with increments to keep up with inflation will slow this exodus.

The active forces are at least meeting their numerical goals. The same cannot be said for the army national guard, the army reserve, and the individual ready reserve (IRR). These reserve units and individuals are essential in any substantial U.S. force commitment to provide the combat, combat support, combat service support, and

Table 5-1. Shortages of Noncommissioned and Petty Officers (E5-E9), End of Fiscal Year 1981.

	Needed	Budgeted	Actual
U.S. Army	266,530	261,061	262,672
U.S. Navy	213,040	191,828	187,345
U.S. Marine Corps	55,460	47,547	50,202
U.S. Air Force	221,222	200,445	200,149

individual replacements for sustained combat. Strenuous efforts of the services appear to be making some headway in the reduction of personnel shortages in the reserve forces. Any immediate or near future crisis requiring mobilization would be extremely dangerous to the United States and its allies, depending as we do upon transporting U.S. forces overseas to honor treaty commitments or to stabilize crisis situations. See Table 5-2, which compares reserve force and IRR authorized strength with actual strength.

Air national guard and reserve units have been maintained at satisfactory levels, but shortages of selected skills are apparent. Strain is introduced with new systems or aircraft. Some of the more demanding missions assigned to regular air force units, such as all weather operations requiring maintenance of complex sensor equipment, are currently beyond the capability of most guard and reserve units. As weapon systems become more intricate in order to meet the demands imposed by improvements in Soviet air units, the personnel of those units must become increasingly capable.

The problem of readiness is obviously complex. The all volunteer force so far has not provided an adequate answer. Additional incentives will be costly, leaving still less than the roughly 50 percent of defense costs for equipment that is now the case. No single solution, whether conscription or all volunteer force, is likely to provide the answer, but some combination of the various proposals, including compulsory as well as voluntary service, seems indispensable.[1]

Training facilities in the United States are generally adequate; the new Army National Training Center at Ft. Irwin, California, affords a new dimension for battlefield training for battalion combat teams. But problems remain. Strict environmental controls at most domestic and overseas installations restrict maneuver considerably. The new weapons systems are of longer range, requiring larger areas for the firing practice so essential to combat effectiveness.

Inflation of fuel costs has imposed financial hardships upon training budgets, in effect decreasing maneuver opportunities for the army. In the case of the air force and the navy, increased petroleum product (POL) costs have affected virtually all programs as money is diverted to meet the several hundred million dollars operations and maintenance shortfall each year. U.S. air force pilots have been receiving about 12 hours per month of flying, although all agree they need about 20. This has reduced levels of proficiency with a resulting impact on combat readiness.

Table 5-2. Shortages of U.S. Reserve Forces, End of Fiscal Year 1981.

	Authorized 1982 Wartime Structure Requirements	Actual 1982 End Strengths Budgeted and Funded	Selected Reserve	
			Authorized	Actual
Army National Guard	446,100	397,651	385,776	378,543
Army Reserve	285,800	242,132	219,538	215,648
Naval Reserve	120,600	87,559	87,400	88,041
Marine Reserve	42,000	38,540	36,653	35,367
Air National Guard	100,800	99,100	98,083	97,518
Air Force Reserve	52,500	63,965	60,754	59,314
			888,204	874,431

Estimated Individual Ready Reserve

Army 250,000
Navy 20,000
Air Force 60,000

Strenuous efforts are underway within the Defense Department to overcome these problems. The abortive attempt to rescue the hostages in Iran in 1980 evoked concern at deficiencies in readiness and training, and the Department of Defense budgets since reflect that concern in marked increases in readiness funding for all four armed services.

Chemical Warfare (CW) Defensive Capabilities

There is little question that the Warsaw Pact has the capability for a chemical offense; NATO efforts are aimed at defending against it. Despite some progress NATO in general lags behind Warsaw Pact capabilities in nearly all aspects of CW defense: individual and unit protection, detection, decontamination, antidotes, and so on.

U.S. policy with regard to chemical warfare has stressed the defensive aspects, resulting in slow and limited development of new weapons for employment against enemy forces. This imbalance is of increasing concern to U.S. military planners in light of loss of U.S. nuclear escalatory control. They feel deterrence against Soviet use of these weapons can only be achieved by a credible capability on the part of our forces to pose a real CW response in kind to the enemy. The United States should expedite the development and procurement of the modern and much safer binary weapons program as well as the procurement and construction of protective measures.

The Forces in NATO

The United States, the Federal Republic of Germany, Canada, the Netherlands, Belgium, the United Kingdom, and France maintain armed forces on the central front in Europe. All except France, which though a member of NATO is not a member of the military command, are responsible for deployment close to the borders with East Germany and Czechoslovakia. See the maps (Figures 4-1 and 4-2) in Chapter 4 showing peacetime stationing of NATO and Soviet ground forces.

Readiness of the NATO central front forces depends upon the size, quality, equipment, training, tactical mobility, and sustainability of those forces. The U.S. forces in Germany number approximately 200,000 uniformed army and 65,000 air force personnel (some in Britain), stationed largely in NATO's central army group (CENTAG) area in the southern half of the Federal Republic. One brigade of army troops is stationed in northern Germany, near Bremen. (See Figure 4-1).

The U.S. army forces are organized in two corps of two divisions each plus two armored cavalry regiments and combat support and service support units and three separate brigades. The U.S. fifth and seventh corps commanders report in wartime and for exercise and planning purposes in peacetime to the NATO CENTAG headquarters, as do the second and third German corps and the fourth Canadian mechanized brigade group. Normally all forces except air defense report to their national authorities in peacetime. Logistics too is a national responsibility. (See Figure 5-1).

The U.S. air force components are likewise stationed in southern Germany and report to the NATO fourth allied tactical air force (4ATAF) in war and exercises, as do the German and Canadian air units. The 4ATAF has peacetime responsibility for air defense. The areas of CENTAG and 4ATAF are identical.

Both CENTAG and 4ATAF report to the headquarters allied forces center (AFCENT), which commands all allied forces in the central region during wartime. The 4ATAF reports through the commander of the allied air forces for central Europe (AAFCE) along with its counterpart in the north, 2ATAF.

U.S. divisions and corps on the central front are maintained at close to 100 percent of peacetime authorizes strength at all times. Replacements are trained in the U.S. training centers before reporting to their units. Although total overall numbers are high, the authorized percentages for noncommissioned and specialist personnel are considerably underfilled. These shortages reflect U.S. military shortages worldwide.

Personnel turbulence also takes its toll. Officers, noncommissioned officers, and enlisted soldiers, who are authorized to bring their families overseas, serve a 3-year tour of duty in Germany. Junior enlisted soldiers serve only 18 months (as of October 1980) largely because of the sociological hardships associated with overseas duty, such as barracks life in a land with different culture and different language. Team readiness at the tank, infantry, and artillery crew level suffers as a result.

READINESS, REINFORCEMENT, AND RESUPPLY   143

Figure 5-1. NATO Command Structure.

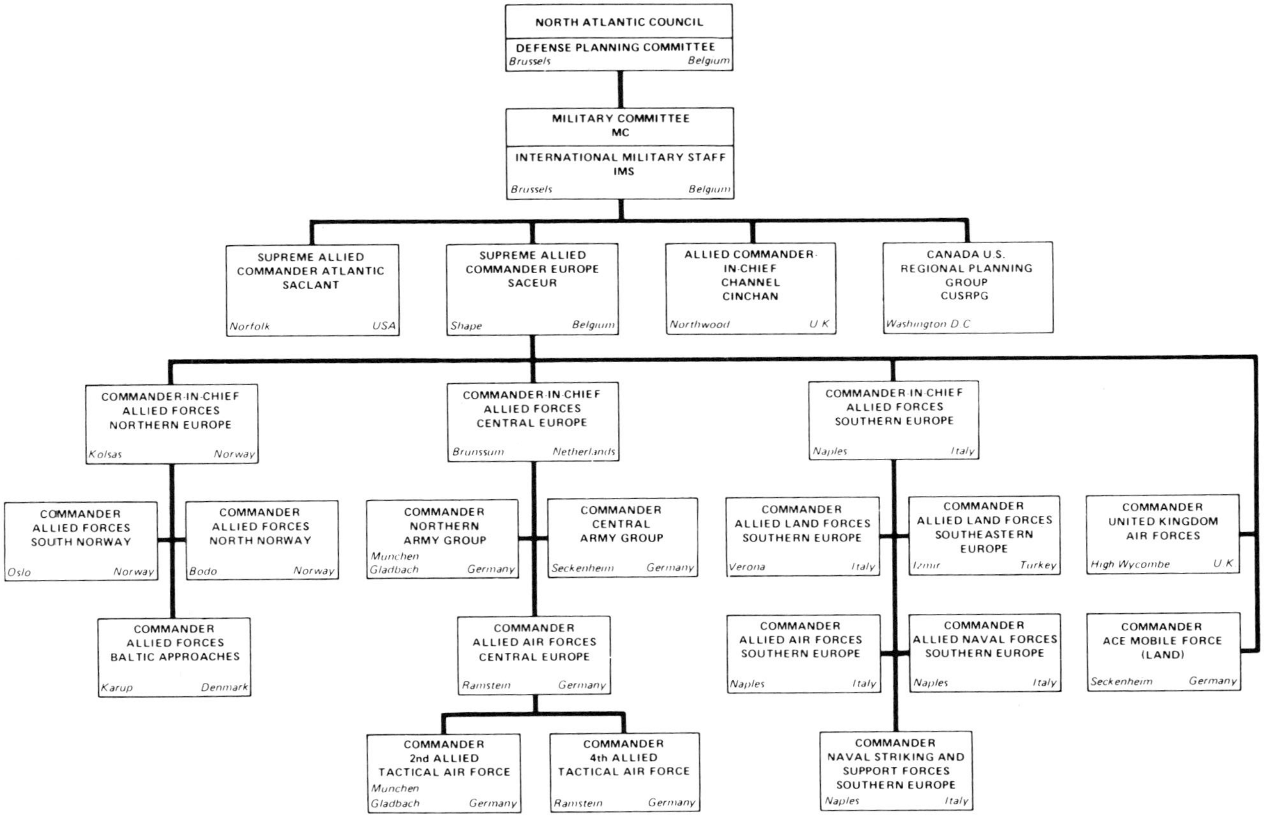

Training is complicated and difficult with the obvious limitations inherent in a crowded, industrialized Germany. Although some exccllent ranges and centralized training areas do exist, the numbers are insufficient to accommodate the large number of units, and commanders must make do during much of the training year with local training facilities, which in many instances are inadequate. This last observation is particularly significant for the future because new weapons and equipment require more, not fewer training areas and greater range and maneuver area. Air units are constrained in the number and size of weapons they can employ on German ranges. Dense commercial air traffic and congestion in base areas also impose restrictions on air training operations. One area of particular concern is low-altitude training. NATO pilots count heavily on being able to fly under enemy surface-to-air missile (SAM) and radar defenses. Altitude restrictions on flying in German air space make much training unrealistic.

Amounts and quality of U.S. army equipment in the deployed forces reflect the same high priority as personnel. Troops are armed and equipped with the latest weapons and equipment available, and in the highly mechanized divisions and corps uniformly high rates of readiness result when spare parts are available over the long supply lines from the United States. Shipping time for spare parts from requisition to receipt is about 135 days by ship and 60 days by air. Equipment on hand in deployed service support units is not at as high a level as that in combat and combat support units. Shortages, particularly in trucks, POL, communication equipment, and materiel handling gear are significant. Unfortunately, budgetary realities restrict the flow of needed spares, oil, and lubricants. Inflation in costs of POL and repair parts, which must be paid for by the overseas commander from operation and maintenance funds, has required a decrease in the amount of maneuver training. The slack is taken up partially by maximum use of local training areas and the increasing proliferation of simulator training devices, but further cuts in the operations budget may result in measurably decreased readiness. Some contend that this decrease has already taken place.

Still another equipment problem is being faced by the army in Washington and Europe. Modernization presently underway will place in the field some 400 new systems before 1990. The modernization efforts to upgrade the army, largely post-Vietnam, have concentrated on mechanized field environment. The new MI tank and the infantry fighting vehicle are examples, but among the 400 new systems are many smaller ones such as communication, automated data processing (ADP), intelligence, and maintenance. For the field army in Germany, normal budgetary problems are compounded by lack of adequate real estate, construction requirements, maintenance, and coordination of training and equipment. It is of the utmost importance that rapid resolution of the modernization problems take place to upgrade the combat effectiveness of the forces.

The concept of storing division and corps sets of equipment in Europe to minimize the airlift load during reinforcement (POMCUS—prepositioned material configured to unit sets) has been implemented for approximately three division sets in CENTAG and action is in progress to store an additional three division sets plus supporting units for NORTHAG. Completion of the ongoing program in the 1980s will reduce significantly the time lag from alert in the United States to reinforcement readiness for combat.

Unfortunately POMCUS priorities conflict with overseas storage of equipment for war reserve materiel, equipment and supplies (including ammunition) for replacement and replenishment for combat forces. Both are essential, but until decisions are made to upgrade prepositioned war reserve materiel stocks, U.S. forces face high risks in sustainability over time.

U.S. air force units in Europe (USAFE) have enjoyed high priority since termination of the Vietnam War. Accordingly they possess the largest percentage of new aircraft and weapons, the largest stocks of both air-to-air missiles (Sparrows and Sidewinders) and air-to-ground missiles (Mavericks), for example. European stockpiles would have to be depleted if the United States were compelled to fight elsewhere, say, in the Middle East.

Questions have been raised about the adequacy of forces planned for Europe in the future. The USAF has been unable to reach the goal of twenty-six tactical fighter wings planned for 1981. That goal will not be achieved before 1985 under current programs. Badly needed systems to increase all weather capability in the demanding weather environment of Europe have repeatedly slipped. The F–16, which will be our major fighter in Europe, still cannot fire all weather missiles. Systems such as LANTIRN, designed to provide night and under-weather missile delivery capability to the F–16, have not been funded sufficiently.

Other consumables, such as spare parts and new fragmentation weapons needed to deal with Soviet armor, have not been stockpiled in sufficient quantities to meet required sustained combat levels.

Although emphasis is now being placed on these deficiencies, it will take time to improve the situation. Despite these problems the modernization of U.S. air units in Europe is taking place at a substantial pace. The introduction of the F-15, the world's most advanced all weather, air superiority fighter, has materially enhanced NATO's air defense picture. These aircraft are based at Bitburg Air Base in Germany and at Camp New Amsterdam in the Netherlands, thus providing coverage of both fourth and second ATAF areas. When teamed with airborne warning and control aircraft (AWACS) scheduled to come into the region in the next several years, a capability to deal effectively with new Soviet attack aircraft such as the Fencer will be at hand. The two F-111 wings based in England provide a long-range, all weather attack capability, which greatly enhances NATO's offensive punch.

Other modernization programs appear more questionable. The F-16 soon to be introduced in USAFE, while constituting an excellent clear air mass capability, will be lacking the marginal weather attributes of the F-4s they will be replacing until they are equipped with the new AAMRAM radar missile. Increased emphasis on improvements to rectify these deficiencies will be required.

The A-10 attack aircraft now assigned to the Royal Air Force (RAF) base Bentwaters/Woodbridge in England and scheduled to operate from forward operating locations in Germany are assigned the primary missions of close army support, including the armor-killing role. Whether they will be able to operate in the heavy surface-to-air missile and ZJU 23/24 defense environment of the central region is a matter of some concern. Tactics to enhance their survivability are being stressed.

Equipment readiness presents different problems. In the U.S. army lack of dollars has meant a significant slowdown in modernization and a long list of new items ready for production pending funds. The result again, regardless of dollar influx, is a period of 5-10 years of risk until modern equipment can be emplaced. It is worth noting that in a number of instances our allies, particularly the Germans, have more modern army items in the field than does the United States.

Quality of equipment of NATO versus the Warsaw Pact deserves special comment. NATO has for years perceived itself to be militarily secure because of the superior quality of its equipment. The Warsaw Pact forces had tremendous numbers of armored vehicles and aircraft, but NATO had the advantage of quality. If that superiority was ever true, it is not so now. Even NATO's vaunted air power, the quality of its sophisticated fighters, for example, is challenged, and the Soviet Union is building two fighters for every one built by the West. The answer to technological superiority in weapons systems lies within the state of the art in the West today in computer-based microelectronics and microprocessing, antitank and antiarmor systems employing precision-guided missiles, millimeter wave and focal plane array technology, better weapons and avionics. Many systems are on the drawingboard today, but they take time to develop and field and they are expensive.

REINFORCEMENT

The concept of unit reinforcement calls for movement of army and air force units to Europe during a crisis and preferably prior to the outbreak of hostilities. Army reinforcement units in echelons to be deployed early are primarily armor and mechanized brigades and divisions, which depart their home bases to joint up with their pre-positioned materiel configured in unit sets (POMCUS), move to assembly areas and prepare for NATO battle commitment. Assuming availability of air transportation, alerted POMCUS forces can be in position in their assembly areas in Europe in approximately 10 days to 2 weeks after notification. It should be noted that aircraft discharging units in Europe for pick up of POMCUS are planned for use as evacuation means for U.S. civilian personnel, station in Europe and dependents. Air force units can move even more rapidly by self-deployment of squadrons, assuming air refueling availability and military airlift command (MAC) transport for equipment that cannot be carried by the unit. But this assumption becomes of questionable validity for both army and air force reinforcement plans in light of airlift limitations and possible interdiction of NATO airfields.

Competition for airlift will be tough even with strengthened civilian reserve air fleet (CRAF) assets. Generally speaking, POMCUS units travel light, the typical army organization carrying primarily individual weapons and light equipment. Problems arise in POMCUS units when, as in a number of cases, only some 90 percent of total unit equipment is in storage in the European warehouses. Some equipment, in particular rolling stock (trucks, POL tankers, forklifts) plus army aircraft and signal gear not authorized for POMCUS are not stored and must be moved with the units, adding to the airlift requirements.

The situation becomes increasingly difficult for the reinforcing forces if combat is initiated prior to their movement overseas. Degradation caused by combat movement losses and possible interdiction of aerial ports of embarkation and debarkation would add to problems discussed previously.

In situations where early warning is acted upon, U.S. air units from the tactical air command, the air national guard and the air force reserve would begin movement to Europe. These units have been receiving theater familiarization through squadron-size deployments for a number of years.

Some seventy-three colocated operating bases (COB) are being readied for their use. In most cases, however, preparations are far from adequate to support sustained operations under combat conditions. It will be a matter of months and even years before the many problems of logistics, munitions, storage, and command and control are all worked out, although these arrangements are being developed under the terms of memorandum of understanding (MOU). Eventually aircraft from all NATO airforces will be refueled and rearmed from a compatible base structure throughout NATO, but this is years away. The lack of interoperability still constitutes a major problem for NATO airforces.

Enemy action of course could have a major impact on air reinforcements. The heavy lift aircraft of Military Airlift Command (MAC), the C-141s and C-5s, are limited to a relatively small number of bases in Europe if they are to enjoy full reception facilities. Should these bases be under attack before air unit deployments have been completed, success of the air reinforcement program would be in serious jeopardy.

The movement of supplies and munitions between storage areas and operating bases could also be tenuous under such conditions. Military planners in NATO have not faced up to the fact that roads and railways now carrying the supply load would be under attack and congested with refugee traffic despite plans for refugee control and thus reduced in their capacity at a time when the need for supplies would be vastly increased. Intratheater airlift would be clearly inadequate to cope with army and air force needs under these circumstances.

U.S. forces deployed in NATO's central region have over time been structured with emphasis on combat capability, as opposed to supplies or support and logistics capability. Since logistics is considered a national responsibility devolving upon each member of the alliance, and in view of the fact that NATO's forces have comparatively little standardization of materiel (despite emphasis upon standardization and interoperability), NATO nations, particularly the United States, have looked to the allies for host nation support, particularly in the early phases of combat. This concept calls for assistance from Belgium, the Netherlands, and Germany (and to a lesser degree the United Kingdom) to assist the U.S. forces arriving at seaports and airports to move to their battle positions, to provide materiel and personnel support to clear depots of stored equipment and ammunition, to provide some security in rear areas, and to assist in rapid runway repair. A Europe rich with logistic capabilities in the civilian and military sectors that are not considered essential for the support of their own forces is willing to provide assistance to U.S., Canadian and U.K. forces arriving on the continent, and arrangements are being made to formalize the essential support. Military planners count heavily upon this concept in order that the United States can deploy the maximum number of combat units rather than support equipment such as trucks in the early days of reinforcement. The risks are obvious.

RESUPPLY

Resupply to deployed units in wartime may be by air or by sea as in peacetime. Major differences exist in wartime, however, and because of the long lead time necessary for some items coming by sea and the heavy requirements upon air transport particularly in the early days, pre-positioned war reserve materiel (PWRM) is stored overseas. In general, the number of days of supply stored are calculated on anticipated combat consumption and the length of time until significant quantities of required items can be received from ship resupply. The present NATO standard calls for a minimum of 30 days of resupply items (PWRM) to be stocked in the theater, with the exception of some spare parts that continue to receive priority for shipment by air line of communication (ALOC). Most nations do not meet this low standard, which is only reluctantly accepted by NATO itself. War reserve materiel is much more than spare parts; it consists as well of major items, like tanks and howitzers, which can be damaged or destroyed during battle and require replacement. Ammunition too is PWRM, and in modern war vast quantities are required. Several problems arise from our present situation. First, POMCUS competes with PWRM, and the United States does not have sufficient equipment for all its requirements. POMCUS has priority; hence PWRM suffers from limited funds. Large amounts of army and air force ammunition are stored in Europe as well, but considerable additional supply is needed to attain acceptable levels of the particular types required. U.S. missiles as a separate and indispensable category are nevertheless in short supply in Europe and in the United States as well.

The effect of these shortages is to require the use of badly needed aircraft early in a war to bring large tonnages of logistical supplies of all types to the theater. As if that were not enough, production in the United States of equipment and ammunition is far from sufficient to satisfy POMCUS plus war reserves needs. Fiscal shortages have required postponing some purchases; in other cases new production or reopening of wartime reserve manufacturing plants would be necessary. This means months and even years would elapse before significant quantities would become available in wartime. Ammunition and missile plant reopenings would be essential before significant resupply could be effected. Problems of transport, reception, and distribution under wartime conditions would further impair the resupply flow. Time might be too short to open an expanded supply line; supplies might be too little and too late.

At present the U.S. production base requires long lead times for new defense production. Orders of critical metals such as titanium are backlogged two years or more. Forgings require two to three years lead time. Major subassemblies, such as landing gears, must be ordered three years in advance.

Increases in production of vital weapons such as air-to-air missiles also require months of preparation. Clearly the U.S. military production base is now alarmingly limited and anything beyond a short war would overtax it.

All military services share in the shortages although each is affected differently. Equipment modernization causes compatibility problems between the active force, normally the first to receive new items, and the national guard and reserves, who receive a lower priority of issue. Because the largest percentage of army service support units are in the reserve forces, they may be prepared to repair only the older equipment used in training the reserves rather than the newer combat gear.

All these shortages together plus the force structure imbalance in Europe in favor of combat rather than support units, increase the necessity for early deployment of logistic troops and materiel by air, together with some combination of prepositioning and host nation support. Competition for airlift is obvious and the result is the imperative for host national support.

An additional problem in ammunition and missile supply is caused by insufficient amounts available to other members of the alliance. Combat action could and in command post exercises does quickly exhaust stockpiles and cause requests for distribution to allied formations before the 45-day arrival time of additional U.S. ammunition by ship. Finally, as combat progresses, sealift requirements increase significantly and a considerable proportion consists of support for the allies' civilian populations. It is estimated that this requirement may add up to two-thirds of the total tonnages anticipated.

There has been no specific discussion of warning time in this chapter, for to cover the subject well would take many pages. Briefly, NATO must be prepared to respond militarily no matter whether warning time is 48 hours or 2 months. As NATO's intelligence capabilities and warning indications analyses have improved, 48 hours has come to be the minimum time of warning of impending Warsaw Pact conventional attack the alliance can anticipate. NATO officials anticipated needing 6-10 days to know enough about Warsaw Pact mobilization for attack before such an attack crosses the border.

Once NATO reinforcement starts, forces already on the European continent could be in their battle positions and most European mobilizations sufficiently far advanced to present a credible defense. Reinforcement by air from North America could be in process too using using active, reserve, and civil reserve air fleet aircraft. Evacuation of American dependents and other civilians would be well underway as well. Note that we do not comment upon the important problem of the political decisionmaking as to mobilization of the NATO members' forces. Needless to say, governments would need some time to analyze warning indicators, become familiar with the situation, and convince their legislatures and their citizens generally of the seriousness of the situation.

THE NORTHERN AND SOUTHERN REGIONS

Readiness, reinforcement, and resupply of the northern and southern flanks of NATO present tremendous problems to the alliance for geographical, political, economic, and military reasons. Militarily the problems examined in the description of the central region generally apply on the flanks as well, with the additional handicap of the longer distance from North America affecting both air and sea movement. In the south, Greece and Turkey's long-standing animosities, the absence of internal infrastructures such as exist in the central region, the Cyprus issue, the emergence of an expanded Soviet Mediterranean fleet, the expansion and modernization of Soviet and Warsaw Pact land and air forces, the extent of obsolescence of NATO South equipment, and the general economic malaise have all contributed to the magnitude of the defense problem.

Combining to make the defense of the northern flank difficult, yet tremendously important are the following: the rapid expansion of the Soviet northern fleet and its increased areas of operations, the interjection of major Soviet forces between NATO forward positions in Denmark and Norway and the reinforcement and resupply distances from North America and NATO Europe, the geographical closeness of the USSR, the small sizes of the standing forces of Denmark and Norway (compensated for to some extent by large mobilization forces), the paucity of air defenses and the overall powerful Soviet political–military pressure in the Baltic area. Although the policy of Norway and Denmark stops short of the peacetime stationing of other NATO forces and nuclear weapons on their soil, it does not necessarily preclude preparations for their introduction in time of war. Loss of the northern flank would seriously imperil reinforcement and resupply efforts for the central region.

In the southern region, a combination of events has further complicated the tasks of deterrence and defense. Turkey not only lies across the Bosphorus, it is of key importance for Soviet access to the Middle East and to the Persian Gulf as well as to the Mediterranean. Its economic plight and the weakening of its armed forces over time as U.S. supplies were reduced together with domestic unrest, make the call for NATO's support of this key ally all the more urgent.

Growing Soviet strength in the Mediterranean, potential Soviet bases in Libya and Iraq, and long-range Soviet aircraft combine to make the allies' task all the more difficult at a time when one U.S. carrier group has perforce been redeployed to the Persian Gulf. Greece and Italy, themselves peninsular powers, cannot help being worried. Perennial domestic challenges to continued ties with NATO are hence of greater concern than in the past. The continuing improvement of relations between Greece and Turkey is fundamental to the security of the southern region.

Canadian, British, German, and American armed forces play an important role in the reinforcement and resupply of the northern flank as do most of them and even more importantly France and Italy in the southern flank. Also in the north the United States and Norway have agreed to preposition equipment for U.S. marine corps brigades to ensure reinforcement in the face of Soviet naval forces in the Norwegian Sea. Constant exercises, more realistic understanding by the governments of countries threatened, and greater resource allocations by the nations involved are all required in order to make deterrence credible.

IMPACT ON NATO OF CONTINGENCIES IN THE PERSIAN GULF AREA

The problems of readiness, reinforcement, and resupply in the NATO area discussed thus far are vastly compounded when considered in connection with possible contingencies outside of NATO, especially in the Persian Gulf area. Although it is impossible to foresee the precise circumstances that might require commitments of forces by and possibly other NATO members, it is possible to consider a range of contingencies and thereby assess the scope of the problems.

Threats to the flow of oil to the West and its allies generally fall within three categories: internal threats within the oil-producing nations, threats developing between nations of the area that might affect oil supply, and threats from outside the area.

The Gulf region has already experienced threats in two of these categories: the revolution in Iran and the Iraqi attack on Iran. Neither situation has compelled intervention of the West so far. Since over two-thirds of Persian Gulf oil production comes from the countries of the West Bank of the Gulf, Europe, the United States and Japan have been able to meet their requirements despite the trouble in Iran. A spreading of the conflict to other Gulf producing areas could have serious impact, however, considering that Western Europe depends on Gulf oil for approximately 66 percent of its oil imports, Japan for 75 percent, and the United States for 25 percent.

Because of these facts the United States has recognized the need to improve its capability to intervene with military force if that should be required. Secretary of Defense Casper Weinberger, testifying before the U.S. Congress in March 1981 stated that in view of the very real possibility that we would be called upon to use military force to defend Western interests in the Persian Gulf region, we have placed great emphasis upon developing a credible capability to project power into that area quickly and to sustain such forces in high-intensity conflict.

Despite general awareness that U.S. military intervention in the Persian Gulf area would have an impact on NATO, the extent of that impact is scarcely recognized. Few people realize there has already been a significant impact. The United States is maintaining a naval presence in the Indian Ocean with a force of thirty or more vessels built around two carrier air groups. One of these groups coming from the Atlantic Fleet, SACLANT's capabilities to secure the flanks of NATO are reduced.

To meet the wide range of contingencies that must be dealt with, the U.S. military planners have earmarked forces for a rapid deployment joint task force. Examination of the units involved (as of the summer of 1981) reveals the implications for NATO.

Naval forces comprise three carrier battle groups, two of which are already in the area, as noted. Not previously mentioned are surface force units, some elements of which are earmarked for NATO along with five aerial patrol squadrons, some with similar NATO commitments.

The marine corps commitment to the rapid deployment force (RDF) involves one marine amphibious force. Here again it is certain involvement of such a force in the Persian Gulf area would require substantial revision of plans to provide marine amphibious force support to NATO's northern and southern regions. In this connection the United States has only sixty-four amphibious vessels (less than one division's lift), and they would all be required for the RDF force if a marine amphibious assault were mounted in the Gulf area, though the fiscal year 1981 and 1982 defense budget request does include funds for additional fast sealift ships.

U.S. air force RDF commitments are also sizable. All twelve of the tactical fighter squadrons involved are NATO earmarked as are the two tactical reconnaissance wings. Of greater importance is the fact that fighter units include F-15s and F-111s that are sorely needed to meet the stringent operating conditions in Europe. Even more serious is the impact of U.S. army commitments to the RDF, including such key NATO elements as the 82nd airborne division, the 101st air assault division, the 24th mechanized division, the 6th air cavalry combat brigade, and two Ranger battalions.

Scenarios could vary of course. The actual combat commitment could be as little as one marine amphibious unit (MAU), which consists of about 1,800 men. A unit of this size is positioned periodically in the Gulf area along with supplies and equipment and could be counted on for a speedy response in time of need. The impact on NATO might be negligible except for taking away marine equipment designated to support the NATO flanks.

If the scenario involved a threat of several Soviet divisions, the picture as far as NATO is concerned could change radically. Moving the committed army units alone would tie up virtually all the sealift and airlift capability of the United States for a number of months. Existing plans call for enhancing U.S. sealift capability, but vessel procurement stretches out over several years. The situation is much the same with regard to airlift. The serious deficiencies that already exist in the ability of the United States to meet NATO reinforcement requirements would be compounded in the event of Middle East demands so as to make the problem unmanageable. The present MAC fleet of 77 C-5As and 276 C-141s would be hard pressed to move the 82nd airborne division and a marine brigade to the Persian Gulf in anything less than three weeks. Adding to this the lift requirements of deploying tactical air squadrons makes the impact even worse.

It could be argued that should the United States be involved in the Middle East at the time of a NATO crisis, it could disengage and redirect its efforts from the Persian Gulf to Europe. Such an assumption requires careful scrutiny. Disengaging forces from combat involvements is a far cry from loading forces out of continental U.S. airfields and ports. One could expect substantial delays in the closing time of all major combat elements rerouted to Europe from the Middle East. Once again the success or failure of such a venture would depend a great deal on the particular situation. It would depend on whether there was ample strategic warning of an impending attack in Central Europe, whether NATO acted promptly on this warning, and so on. Such matters require careful analysis by both U.S. and NATO planners.

It would seem apparent that the NATO partners should begin serious discussions now with regard to the inevitable impact on NATO defense plans of U.S. involvement in the Middle East. It is suggested that the following proposals be given careful consideration:

- Create combat units from reserve elements of European NATO members to offset U.S. forces earmarked for rapid deployment.
- Procure equipment sets for such units.
- Create additional combat support and combat service support forces for passage to NATO command.
- Speed up reserve mobilization procedures to enable fast activation of these reserve units. (Swiss and Israeli practices bear examination.)
- Position current forces within NATO for better tactical advantage.
- Improve command and control capabilities of NATO headquarters.
- Improve firepower to compare more favorably with Soviet capabilities.
- Increase naval commitments of European NATO members.

Other items could be added to this list. Most important is that dialogue get underway. Protection of Middle East oil is a crucial part of the overall blueprint for a successful defense of NATO.

The civilian and military leaders of the United States must weigh the extent to which future military involvement in the Middle East would affect vital interests in collective security elsewhere in the world. In view of the potential dilemma in allocating our defense resources, it becomes increasingly clear that hostilities outside the NATO area would gravely increase the risk of a more direct armed confrontation between East and West on a far broader front.

LOGISTIC COMMENTARY

Navy and sea power, too often taken as synonyms, have, in the case of NATO suffered the same difficulty, in that the entire focus of ministers of state and military commanders alike over the life of the alliance has been on warships and not on lift. The totality of sea power embraces the aggregate of warships, merchant ships, aircraft, and bases from which they may operate with confidence. This means that the bases must be available rain or shine seven days a week. Having to count on last minute makeshift and usually highly problematical arrangements is and always has been quite unsatisfactory. Navies have always been very logistic in character, intended to keep the proverbial sea lanes open to ensure the free flow of commerce. This commerce, often taken lightly and all too often overlooked, includes the commerce of peace and profit, and in times of stress quickly transforms to the commerce of combat. As America has become increasingly dependent upon natural resources not found in the United States own land, our appetites have had to be satisfied from overseas. Satisfied they have been, but at a staggering price to national independence and self-sufficiency. Not only have we frittered away our independent capability to move essential materials, raw and in finished form, we have come to see the day when the lift being used is not only not our own, it is not even in the hands of allies.

Since logistics is not a popular topic, in peacetime it always seems necessary to justify and rationalize navies' sizes and even their very existence by calling them by names calculated to inspire confidence in their powers to repel the enemy, like "presence forces," "projection forces," "deterrent forces," or "strike forces." No matter how impressive the name, when all is said and done, the telling mission of a navy is to make sure the soldiers on the battlefield don't run out of beans and bullets before beating some sense into the aggressor's head. NATO today has enough warships to do a barely marginal job on about one-half of the missions that need to be done simultaneously. The alliance has done this to itself through gradual attrition apparently on the assumption that all has been well so far: We have had no great thundering war in Europe and ships are very expensive, so the tendency is to cut a few more from the fleet and see what happens.

At the same time an even more insidious bleeding of the patient, the totality of alliance sea power, has been going on, as the Soviet Union has been allowed to expand its merchant marine using the commercial ploy of loss leaders to garner shipping contracts until now we find the Bear throughout the oceans.

As the USSR has expanded its world trade and hence its sphere of economic influence, it has built up its own navy to give it stability, credibility, and assured freedom of movement without outside interference. Concurrently recognizing the insular position of the United States and the total dependence of NATO on rapid reinforcement and resupply to be credible in Europe, the USSR has built additional maritime tools of the trade with which to confound our logistic efforts. We must keep in mind the Soviet Union can drive to war or take a bus, whether in Europe or in the Middle East, whereas the route for the United States to the Middle East is a water highway some 14,000 miles long via the African Cape. Some would dispute the need to consider the Cape route, preferring the Suez Canal, but the Suez, as we know, is vulnerable to closure. It is easy to close the canal but has taken a year or more to open each time it was closed, and those occasions were benign. Using every last piece of heavy lift equipment in the non-Communist world, it took one year to move ten wrecks from the canal: The heavy lift shifts, cranes, and equipment had to be towed from northern Europe and from the Far East across the Indian Ocean. Their vulnerability en route need not be underlined. The point is that time and distance are ever present factors to be considered along with vulnerability in transit.

On the matter of vulnerability we seem determined to ignore it or perhaps anticipate a breakthrough offering levitation by the time we have to fight. Losses will be staggering in the first few months of any shooting involvement where the opponent has some ships and submarines. The Soviets of course have many fine ships and a surfeit of submarines, as do many other smaller nations whose alignment when the chips go down is speculative at best. When we begin losing valuable merchant bottoms as we undertake to reinforce and resupply in advance of combat on the ground in any theater and we do not know to whom the submarine belongs that did the sinking, with whom do we go to war? It could be the wrong enemy. Some self-serving rascal just might have it in mind to force our involvement and then stand aside while the fur flies. The only acceptable answer is to provide adequate protection en route all the way and make it clear at the start we will tolerate no unidentified predators close to our ships and airlift in transit. This means big numbers of ships to do the escorting. Committing big enough numbers to the escort role in theaters beyond the NATO boundaries immediately emasculates the alliance's navies needed numbers in the alliance boundary area to strengthen Europe itself—Hobson's choice once again.

It took between ten and fourteen destroyers plus a capital ship as ocean escort to match possible large raiders such as a Bismarck in World War II before attrition was overcome. We still turn our backs on such loss potentials as being alarmist, unreal, and unreasonable; yet they are fact, they happened, and can and probably will happen again perhaps to a more aggravated degree because the range and accuracy of weapons have much improved. Over sixty merchant ships sailed toward Malta in one 8-month period during World War II; fewer than ten arrived intact and unscathed; most were sunk. To do that well it cost four carriers brutalized, one sunk along with some twenty destroyers and cruisers smashed and sunk. Malta survived, but just barely. The run from Diego Garcia to the head of the Persian Gulf is a longer haul than the run in the Mediterranean to Malta. Testimony in hearings and public statements does not indicate there are any spare bottoms or duplicate sets of equipment in Diego Garcia either today or planned for the future. Neither the Vietnam nor the Korean experiences may be used as reference points because a benign environment existed then fortunately, as far as long-haul sea and airlift were concerned.

The last time the United States undertook a far distant campaign in a no-holds-barred environment was Okinawa. There the distance was but 7000 to 8,000 miles (not 14,000), and we had four years to work up to the task. We anticipated an enemy 65,000 strong without the remotest possibility of their being reinforced. Intelligence was off by 20,000; the Japanese had 85,000 in place on the island. This knocked our timetable into a cocked hat; it took months longer than anticipated and over 185,000 troops to do the job. To get them there and ashore and to support them took over 1,300 ships, merchant and naval. Today we have only 40 ships of all types in the Indian Ocean, and reports indicate it will take years to build up the capacity to support three brigades.

Aside from the devastating impact on the NATO theater reserves any heavy commitment in a far off theater would have, there is the matter of time and distance to be addressed to withdraw, reconstitute, and travel back to NATO areas. To bring such magic to the stage would require an abundance of dedicated airlift, sealift, and a particularly cooperative opponent who would let you withdraw gracefully and with enough intact at the end of the exercise to have made the exercise worthwhile in the first place. Germany tried it when the decision was made to bring Rommel's Afrika corps, or what was left of it, the few short miles across the Mediterranean to Europe. Their attempt was a disaster. More recently we have the uninspirational record of what had to be left behind in Vietnam.

At least one-third of the Atlantic available naval power would have to be tied down to support a far flung Indian Ocean drill. Roughly one-half of what would be needed is available and operating on any given day in the aggregate of all the NATO navies. Reinforcement from the Pacific has been dreamt of for years. The so-called swing strategy would merit its name if we could count upon large force shifts but in reality the only thing swinging at the end would be the interests of the United States and of the NATO alliance—at the end of a Soviet noose. If NATO ever has to fight, the battle will be global in scope, devoid of any sanctuaries, and with forces tied down at the outset, fighting in some far off places and thus very probably unavailable for rapid redeployment to NATO Europe for months at a minimum, and probably longer. It is no facile matter to disengage quickly, move thousands of miles, and be ready to do battle in a new clime with the same level of effectiveness as coming to action for the first time from a bedded down posture.

In terms of other maritime forces, the merchant ships providing the lift for pre-positioning, reinforcement, and resupply, the fact that the United States is finding it necessary to buy, lease, and charter the needed mundane logistic lift for the modest commitment in the Indian Ocean from foreign navies and foreign merchant fleets tells the story in painfully embarrassing detail. This is indeed a national mortification. These highly specialized merchant ships are few in number, expensive to operate, but absolutely essential to the credible deterrent posture it is mandatory we project. The navies of the alliance have quite enough on their plate to meet shortfalls in naval ships. Navies have understandably assumed the maritime administrations of nations would do their jobs, but they have not been allowed to simply because it would have been too costly to subsidize or otherwise provide a reserve of requisite lift of the proper speed, numbers, and configuration. NATO's civil emergency planning committee has been a voice in the wilderness for years on this score—

a voice, nothing more. Needs have been known, documented, put forward, and then set aside as too costly. The "short war" syndrome has emerged as the lazy man's way out, the easy alternative: "Let's go nuclear early and save all that money for logistic wherewithal." But where are we going to spend all this money we have been saving in the devastated mess of a nuclear aftermath.

The foregoing observations are dangerously incomplete without mention of NATO's potential for a surge of war production and the prior step of stocking inventories with which to fill the sealift and airlift while the industrial base is getting up to speed. That potential is inadequate. As in conflicts in the past, after a few months in some cases, weeks in other cases, and days in a few others, forces will have reverted to much older but more numerous weapons where the consumables are more abundant and more easily and quickly turned out. Such alternatives have not been well addressed by any nation. Here again is evidence of the euphoric expectation that the problem will go away or, it is hoped, never come up.

Enough sealift is owned by the NATO nations together to do the job, but some 25 percent is under flags of convenience. The mix of ship types is not optimal. It needs to be improved, but the needed types for military deterrence and credibility are not optimal for commercial competitive use. The same applies for airlift except that the numbers of available aircraft, even with the mobilization of the civil reserve air fleets, is far from sufficient to meet demands just for NATO, much less in some skirmish designed by an enemy to tie down known NATO forces in some distant corner of the world. NATO nations own some 10,000 oceangoing vessels. Europe and nations on the flanks of NATO will require some 6,000 shiploads arriving each month from North America, the Far East, Africa, and elsewhere to reinforce and resupply in the event of war. Most of those ships would be carrying cargoes for the subsistence of civilian populations at one-half of peacetime consumption rates. That means half the food and half the energy and half of everything else delivered today. Existing naval power in the alliance could offer to those 6,000 ships roughly one-half of the protection they need to make the trips at risk levels commensurate with the end of World War II. The losses promise to be staggering at the outset. Missions and tasks will have to be listed according to priority, which means that someone, some commander, some region would be last, and that is a position no one wants. If priorities become complicated by the drain of force and lift to some remote area, in-place forces and populations suffer accordingly. They would have to hold alone for longer, perhaps for many months, before anticipated commitments could be mustered and transported. Equally unacceptable though is the alternative: focusing solely upon NATO Europe to detriment of the rest of the world and even then being able to do only 50 percent of the job on time.

Rapidly redressing these imbalances remains the only reasonable option. Resolve is only clearly evident by way of an unmistakable posture of staying power. Logistic adequacy takes its place as an element of strategic deterrence the equal of or greater than any nuclear weapon system, newly outfitted divisions of troops, wings of aircraft, or surface-to-surface ballistic missiles (SSBNs) on station.

The United States and all of the NATO partners must together recognize that logistics is indeed a strategic requirement. National efforts are not strategically oriented but reflect at best national priorities and, even more, simple neglect. Responsible leaders must come to understand that this lack of coordinated logistical efforts imperils the NATO goals to which they are dedicated. And national legislatures, including the U.S. Congress, must restore priority to the fulfillment of what are national responsibilities.

The cumulative effect of a decade or so of a strength-sapping inadequate defense effort means that a comparably extended period of recuperation cannot be avoided. That need not be cause for despair. Encouragement can be garnered from the likelihood that circumstances are likely to afford us the time to restore a stabler balance of forces so long as the allies, jointly and severally now embark unmistakenly along the path of a sustained effort to remedy the situation. Demonstrating the intention to persevere in creating an adequate deterrent and defense is sorely needed from every point of view political as well as military. Conversely a failure to check and revise this past trend could well carry the current and impending imbalance of forces beyond the danger point.

NOTES TO CHAPTER 5

1. The Atlantic Council has recently commenced an in-depth study of the military personnel problem under the co-chairmanship of Dr. Lloyd Elliot and General Andrew J. Goodpaster.

6 THE IMPACT ON NATO OF SECURITY REQUIREMENTS OUTSIDE THE TREATY AREA

Jeffrey Record

The West today faces a grave security crisis outside the NATO area. Certain and uninterrupted access to resources vital to the West's economic survival is being jeopardized by myriad threats challenging not just Western economic interests but the West's very capacity to respond in a coordinated and effective fashion.

Events of the past decade in Africa, the Middle East, and Southwest Asia raise serious doubts about the continued utility of the North Atlantic Treaty Organization as the primary instrument for the collective defense of Western security interests. Both the locus and character of the main threat to those interests have radically changed since the alliance was formed in 1949. As former Secretary of Defense James R. Schlesinger observed, in 1949

> we needed to protect the land mass of Western Europe against the possibility of Soviet invasion and to provide for the recovery of Western Europe. Those requirements were met through the creation of NATO and the Marshall Plan. Today, the security problem has taken on an altered form. The easiest route to the domination of Western Europe by the Soviet Union is through the Persian Gulf. And it is to be noted that NATO is a defensive alliance. It cannot in terms of its own charter respond to what may be the more serious threat against security of Western Europe.[1]

On the contrary, during the past decade it has become apparent that NATO's very success in deterring a direct Soviet military advance against Western Europe may have encouraged Moscow to adopt an alternative, pseudomilitary, indirect approach to undermining Western security. Blocked in Europe, emboldened by its rising comparative military advantage, and encouraged by the flaccidity of Western responses to it, the Soviet Union may now be seeking to gain a stranglehold on the economic foundations of Western security. That Soviet global intentions may encompass a flanking maneuver across the West's vital but comparatively undefended economic underbelly in the Third World is certainly not belied by Soviet and Soviet-sponsored violence in Angola, Mozambique, Ethiopia, Yemen, Iran, Afghanistan, and Indochina. It may account for the spectacular growth in the size and capabilities of Soviet surface naval, amphibious assault, airlift, airborne, and other forces dedicated to the projection of power beyond the traditional confines of the Eurasian land mass, and for the establishment of Soviet-controlled military bases astride the West's economic lines of communication with the Third World.[2]

Soviet activity in the Third World not only poses a potential threat of economic interdiction; forces capable of denying the West access to the raw materials of the Third World are also forces capable of disrupting the movement of vital U.S. reinforcements across the Atlantic in the event of war in Europe. That this new challenge to Western security outside the NATO area constitutes a threat as potentially deadly as any within the NATO area is glaringly evident in a host of statistics on the present and projected Western dependence on Third World oil and other critical raw materials. The sensitivity of the West's economy even to momentary disruptions in the steady flow of Third World raw materials was graphically demonstrated during the oil embargo Organization of Petroleum Exporting Countries (OPEC) of 1973-74. Less well appreciated was the embargo's direct and immediate impact on NATO's military readiness. Within a few weeks the U.S. navy had to reduce steaming hours by 20 percent, while the U.S. air force cut flying time by 33 percent.[3] In Europe NATO training exercises were sharply abridged and fuel-sharing arrangements undertaken among various national commands.

It is important to keep in mind that Western economic dependency on the non-Western world is not a new phenomenon. Since the beginning of the industrial era, Europe has been increasingly reliant on access to raw materials outside itself. The present geostrategic crisis confronting the West is thus not attributable to the need for assured access but rather to the fact that access is no longer assured.

For well over a century, uninhibited Western passage to the fossil fuels and mineral resources of Africa, Asia, and the Middle East was guaranteed through the medium of colonial empires and by continued, unchallenged Western control of the seas in the immediate postcolonial era. By the late 1970s, however, three seminal developments had transpired that placed that access in jeopardy. The first was the collapse of Europe's colonial empires in the 1940s and 1950s and the emergence of a host of largely unstable and often warring states incapable of providing the requisite political stability associated with the West's traditionally untrammeled access to vital non-Western sources of raw materials. The second was the steady recession of Western military power outside the North Atlantic area that accompanied the demise of colonialism, highlighted by the withdrawal of British forces east of Suez in the early 1970s. The third and most ominous development has been the relentless Soviet attempts to establish its own military power in critical areas thus vacated by the West.

What for the West was once a purely economic matter has become a security crisis as well: What were always vital economic interests whose security was taken for granted are for the first time being challenged by a combination of local instability and the willingness of a hostile power to exploit it to the West's disadvantage. The geostrategic crisis, stemming largely from the existence of vital interests unattended by a military presence and capability sufficient to deter and defeat threats to those interests, is compounded by several factors. Unlike the Eastern bloc, which is dominated by a single state capable of orchestrating a coordinated international threat (combining Soviet weapons, East German technical advisors, and Cuban troops for example) to Western interests outside the NATO area, NATO is a grouping of sovereign states possessing disparate and sometimes conflicting national interests in the Third World. Indeed, their mutual military obligations are confined by treaty exclusively to the North Atlantic area despite the presence of shared interests outside that area now being threatened by the same adversary whose behavior in Europe in the 1940s sparked the writing of that treaty. As an institution NATO is ill-suited for dealing with military challenges outside Europe.

The level of dependency on Third World raw materials differs considerably among the member states of NATO, leading to a divergence in perceptions of the threat, proposed responses to it, and levels of vigor with which those responses are pursued. This is particularly the case with respect to Persian Gulf oil: for example, 61 percent of Western Europe's imported oil flows through the Strait of Hormuz, compared to less than 10 percent for the United States.

The far greater reliance of Western Europe and Japan on Persian Gulf oil has contributed to our allies' perceptions of the Soviet threat to the region and proposed responses to it being quite different from those harbored by the United States. The United States, whose economy is less vulnerable than Europe's or Japan's to an interruption in the flow of Middle East oil, has (at least since the Soviet invasion of Afghanistan in December 1979) been more forceful than its NATO or Asian allies in responding to Soviet penetration of the Middle East and Southwest Asia. In contrast, Japan and the nations of Western Europe, seemingly numbed by their utter dependence on Persian Gulf oil, have yet, as Albert Wohlstetter has correctly observed, "to face candidly the dangers made visible by the recent crisis in the gulf region."[4] This different perception of the threat has been manifest in the lukewarm Japanese and European support provided for strong countermeasures proposed by the United States. The U.S. imposition of a partial embargo on grain sales and technology transfers to the Soviet Union in the wake of the Afghan invasion was not accompanied by any significant similar measures on the part of its allies. Nor did the allies respond adequately to President Carter's call for a boycott of the Moscow Olympic games: France refused to join the boycott; Japan did so only under U.S. pressure; and Germany postponed its decision for fear that the administration would change its mind.

In fairness it must be admitted that more is involved in the allies' reluctance to follow the U.S. lead than simply varying perceptions of the threat. As *New York Times* journalist Richard Burt has astutely noted,

> The disinclination . . . of some European governments to adopt a tougher line toward Moscow not only reflects fears of possibly entering a new Cold War. It also reveals the new caution of Europeans in following American policy initiatives. The caution, obviously, is not difficult to understand: for three years America's allies in Europe have been almost continuously surprised and irritated by Carter Administration policies that they neither expected nor understood.
>
> The confusion and inconsistency of Carter administration policy toward the Soviet Union is a case in point. While highly critical of Moscow's human rights performance at home, the administration only gradually grew con-

cerned with the projection of Soviet power abroad. Rapid reversals in policies toward Moscow led some European governments to conclude that Washington's strong initial reaction to Afghanistan would probably soon be replaced with a more conciliatory line. But European confusion over Washington's policy toward Moscow is only part of the problem. Other aspects of American foreign policy over the last three years, particularly efforts to stop the proliferation of nuclear weapons, technology, and conventional arms, have also irritated Europeans and contributed to the fundamental divergence of perspective that now threatens NATO.[5]

A third and closely related factor compounding the West's geostrategic crisis is Western Europe's sense of greater dependence on Soviet goodwill than exists in the United States. If Western Europe has yet to face candidly the dangers to its security posed by Soviet expansion in the Third World, it seems much more fearful than the United States of the potential dangers in challenging that expansion. The proximity of massive Soviet military forces to NATO Europe; Western Europe's heavy investment in trade with the Soviet Union, including a rising dependence on Soviet natural gas; the reliance of many allied governments on parliamentary support from parties and factions ideologically committed to a pervasive and permanent politicomilitary détente with the Soviet Union—all serve as dampers on Western Europe's willingness to challenge surrogate or even direct Soviet aggression outside Europe.

Even within Europe continental members of NATO have manifested a comparatively greater reluctance to undertake actions that might appear provocative to Moscow, the latest example being the lukewarm and still hesitant support commanded in Europe by NATO's long-range theater force modernization program (LRTNF). Europe and particularly the Federal Republic of Germany admittedly continue to enjoy a disproportionate share of whatever benefits remain of East–West détente, a process that was begun in the late 1960s and that stemmed largely from initiatives undertaken by then Chancellor Willy Brandt. This relatively greater investment in détente with the Soviet Union is evident in policy differences with the United States (especially the Reagan administration) over the role and future of arms control negotiations with the East, and in Europe's predilection for what might be termed differential détente: a willingness to distinguish Soviet behavior outside Europe from Soviet behavior inside Europe as a means of preserving Europe as an island of détente. Soviet actions that would be unacceptable in Europe and might even lead to war are tolerated and often ignored outside the NATO area, even though those actions may pose a no less dangerous, albeit indirect, threat to Western security.

Unfortunately the geostrategic crisis confronting the West today is not divisible along neat geographical boundaries. Can the West meet the burgeoning Soviet challenge to its interests outside the NATO area if it chooses to apply two separate standards for judging the Soviet Union's international behavior? Or to put the question in broad operational terms: Can the United States be expected to sustain a policy of confrontation with the Soviet Union in the Persian Gulf while the Europeans continue a policy of cooperation with Moscow in Europe? Simply to pose the question is to answer it.

A final factor is the probability that the Soviet Union itself will become a net importer of oil. Although predictions vary as to when rising Soviet requirements for oil (which include supplying Eastern Europe and maintaining a level of oil and natural gas exports to the West sufficient to finance imports of badly needed grain and advanced technology) will exceed domestic production, most analysts agree that the Soviet Union will become a net importer of oil well before the turn of the century. As such, Moscow might have an additional incentive to gain control over Persian Gulf oil supplies; control would provide a stranglehold on the West's economic well-being and a possible solution to what may be an impending *Soviet* energy shortage.

THREATS TO WESTERN SECURITY INTERESTS OUTSIDE THE NATO AREA

A major obstacle to determining appropriate responses to threats to Western security interests outside the NATO area is the diversity of such threats. The spectrum of challenges is so broad as to deny universal utility to any single type of response.

Specific threats to Western security interests in the Third World may be grouped into three categories. The first is the direct employment of Soviet forces against a Third World state. Until December 1979 this threat was dismissed in many Western quarters: For decades the Soviet Union had confined its military conquests to the European area (Mongolia being the sole exception) and had relied exclusively on surrogate forces to achieve power and influence outside Europe.

The Soviet invasion of Afghanistan, however, demonstrated both a capacity and a willingness to employ military power directly against a non-European state in a fashion that could pose a distinct menace to Western interests outside the NATO area. Given the steady deterioration of central authority in neighboring Iran (exacerbated by Iran's ongoing war with Iraq) and a history of unflagging Russian imperial designs on that country, the invasion of Afghanistan threatens to compromise defense of the entire structure of Western interests in the Persian Gulf and Southwest Asia. The establishment of a Soviet Afghanistan almost doubles the length of Soviet-controlled border with Iran, brings Soviet forces some 300 miles closer to the Kremlin's realization of a centuries-old dream of a warm water port in the Arabian Sea, and opens the lengthy Afghanistan–Pakistan border to potential Soviet penetration. From a purely military standpoint the occupation of Afghanistan threatens an already tenuous balance in the region; aside from an impressive demonstration that the Soviets have achieved an operational mastery of their burgeoning capacity to project military force abroad (at least into areas in close proximity to the USSR), the occupation extends the reach of Soviet tactical airpower to areas heretofore regarded as aerial sanctuaries by the West, most notably the Gulf of Oman and the critical Strait of Hormuz. In so doing, the occupation of Afghanistan furthers the erosion of the East–West naval balance in the Indian Ocean.

Whether Soviet forces will advance beyond Afghanistan (against Iran or Pakistan) remains a matter of speculation. In Afghanistan Moscow obviously was deterred neither by Western military power in the region nor by the prospect of indigenous resistance, although the stoutness and resiliency of the latter clearly was not anticipated. Admittedly the logistical burden of sustaining major aggression against Iran or Pakistan would be far heavier than that in Afghanistan; moreover, a Soviet invasion of either country could enflame the entire Islamic world, provoking a level of indigenous armed resistance far exceeding that in Afghanistan. With respect to countries on the other side of the Arabian Sea, the Soviet Union would confront formidable logistical difficulties stemming from the lack of a contiguous force presence (that is, the absence of direct land lines of communication).

Nevertheless it would be imprudent for the West to discount entirely the possibility that another Afghanistan might be visited upon a Persian Gulf state. With the possible exception of Iraq and Iran, no Persian Gulf state possesses military power sufficient to guarantee success against even those forces the Soviet Union could bring to bear against them by sea and air. Such forces include no fewer than seven airborne divisions; each division contains over 300 armored vehicles, more than may be found in either the Omani army or the army of the United Arab Emirates, for example.[6] Indeed, uncertainty over the possibility of another Afghanistan was implicit in the Carter doctrine. Stimulated by the Soviet conquest of Afghanistan, the doctrine was directed against further "outside" aggression in the Persian Gulf and Southwest Asia.

A second category of threats to Western interests outside the NATO area is regional transnational aggression by a Soviet client state. Examples include the Angolan attack on Shaba in 1978, the Vietnamese conquest of Cambodia in 1979, the Iraqi invasion of Iran in 1980, and the Libyan occupation of northern Chad in 1981. Possible future manifestations of this more subtle threat are a Vietnamese invasion of Thailand, an attack on Oman by South Yemen, and an Iraqi move to gain control of Kuwait and the oil-bearing regions of Saudi Arabia.

Notwithstanding the present conflict between Iraq and Iran, the last contingency merits particular attention because it exerts a major influence on U.S. force planning for non–NATO contingencies and because the Soviet-supplied Iraqi armed forces are the largest and most powerful of any in the critical Persian Gulf area. Although Iraq is certainly no puppet of the Soviet Union, that country's foreign policy since 1968 has been distinctly hostile to the West and to the more conservative Arab regimes on the Arabian peninsula (including Kuwait and Saudi Arabia). This hostility constitutes a potentially explosive factor in the region, given the comparative military weakness of Iraq's neighbors. Iraq's active military forces, comprising some 220,000 troops, over 5,300 tanks and armored fighting vehicles, and 310 combat aircraft, exceed the combined ground forces of Saudi Arabia, Oman, North and South Yemen, Qatar, Kuwait, and the United Arab Emirates.[7] More significant, its firepower and tactical mobility dwarf that of the ground forces the United States could rapidly bring to bear on the Arabian peninsula from North America.

The third category of threats to Western interests outside the NATO area is the internal overthrow of regimes friendly to the West, either by Soviet-sponsored subversion or by purely indigenous forces hostile to the West. Examples of Soviet-sponsored internal revolutions are the Neto regime in Angola, the Mengistu regime in Ethiopia, and the Frelimo government in Mozambique. All of these countries are positioned along critical Western maritime lines of communication with the Third World. The establishment of Soviet naval and air bases in those countries similar to those installations already constructed or abuilding at Cam Ranh Bay, Socotra, and Aden constitutes a potential vise on the West's economic jugular.

That a national revolution *not* supported by the Soviet Union can be just as detrimental to Western security interests as those that have taken place in Africa with Soviet and Cuban assistance was graphically demonstrated in Iran. In fact the collapse of the shah and the rise of the Ayatollah Khomeini underlines a central reality that the West cannot afford to ignore in devising responses to its threatened interests outside the NATO area: the turmoil that has engulfed the non-Western world since the demise of colonialism is attributable for the most part to indigenous political, social, and economic factors over which neither the West nor the Soviet Union has any immediate influence, much less control. Although doubtless the Soviet Union has sought to exploit that turmoil for its own purposes and in so doing has contributed to instability, it would be a profound mistake to view Soviet policies in the Third World as the principal cause of instability.

Poverty, corruption, autocratic rule, excessively rapid economic development in a backward social and religious environment, gross maldistribution of wealth, inequitable ownership of land, and ethnic, religious, and tribal antagonisms—these are the ultimate enemies of Western interests in the non-Western world. As such, the problem of *internal* aggression, supported or not by outside forces, is far less susceptible to purely military solutions than the problems of direct Soviet aggression and transnational aggression committed by a Soviet client state. In discussing the question of how best the West should respond to the emerging global threat to its interests, the British defense white paper of 1980 correctly observed that

> The best answer is to try to remove the sources of regional instability which create opportunities for outside intervention. In some circumstances, military measures will not be appropriate at all; in others, they may form only one component of the total response. Diplomacy, development aid and trade policies will usually have a great contribution to make.... Nonetheless many forms of defense assistance can and should play a part in the support of friendly nations.... Over and above this, the West must make it clear to the Soviet Union and its allies that it is capable of protecting essential interests by military means should the need arise.[8]

REQUIREMENTS FOR WESTERN SECURITY OUTSIDE THE NATO AREA

Clearly, the West has vital interests outside the NATO area which no less clearly are being challenged by a combination of indigenous instability and a growing Soviet capacity and willingness to exploit turmoil. To meet the challenge of regaining secure and uninterrupted access to the oil and other Third World raw materials essential to the economic survival of the West, a series of military and nonmilitary measures would appear necessary.

1. A substantial presence of Western military forces must be reestablished outside the NATO area, particularly in the critical Indian Ocean. The steady recession since 1945 of Western military power from the non-Western world, which has created a vacuum of force that the Soviet Union has all too willingly filled, must be reversed. If the West is to deter direct threats to its vital interests in the Third World, it must remarry those interests to a visible presence. Vital interests unattended by military power requisite for their protection provide a standing invitation to hostile adventure.

How large and visible a reestablished Western military force outside the NATO area should be is a matter of continuing debate, at least within the United States. What is not in dispute is the need for a larger standing Western force in the non-Western world. Many believe that such a presence must be largely maritime in character, since there are currently few prospects for the reestablishment of a major Western base ashore in the Third World. The repositioning of Western ground and tactical air forces on the territory of sovereign states that were formerly colonies of the West is unacceptable to most of those states because it would threaten the internal legitimacy of the very regimes the West seeks to preserve.

To be sure, ground forces and land-based tactical air forces would be critical in any major contingency in the Third World (such as a Soviet invasion of Iran); and investment in expanded sea power should not be made at their expense. On the other hand, sea power would be essential in any contingency in areas where the pre-positioning of forces ashore is not possible.

2. The West must create a substantial capacity to project military force ashore in the non-Western world, should its military force offshore fail to deter a direct challenge to its interests. Meeting this

requirement may entail increases in existing amphibious assault capabilities in order to deal with contingencies involving contested entry. It will certainly demand major increases in the strategic mobility of Western ground and other general purpose forces through larger numbers of strategic airlift and sealift platforms and through the concept of maritime pre-positioning. Again, it is not within the purview of this brief chapter to prescribe specific programmatic actions or levels of military investment; however, it is important to recognize that the new geostrategic requirements for power projection capabilities have emerged against a backdrop of over two decades of declining Western investment in such capabilities. For example, the level of U.S. amphibious shipping is at its lowest point since World War II.[9] Moreover, U.S. strategic airlift and sealift capabilities are currently insufficient to meet the stated requirements of a war in Europe, to say nothing of those of a simultaneous conflict in a logistically remote area such as the Arabian peninsula.[10]

3. A third essential requirement for Western security outside NATO, related to both the reestablishment of a standing Western force in the non-Western world and an enhanced capacity to project force ashore, is the restoration of the West's sagging military reputation, particularly that of the United States. By *reputation* is meant a demonstrated willingness to employ force in defense of vital Western interests outside the NATO area and an ability to do so competently.

The decline of Western military power vis-à-vis that of the Soviet Union during the past decade has not gone unnoticed in the Third World, nor have years of U.S. inaction in the face of mounting Soviet-Cuban military penetration of Africa, the Middle East, and Southwest Asia. The Atlantic Council of the United States has correctly observed that

> there has been a change in the perception of Western power and influence on the part of Third World nations. The withdrawal from Vietnam, the Guam doctrine, the Vietnam syndrome as exemplified by Angola, and the decline in Western military capabilities—not only American, but British and other Western capabilities as well—Soviet successes in Ethiopia and Afghanistan, together with the trend toward an overall balance of power favoring the Soviets, all contribute to lessened confidence.[11]

The council might have added that confidence in the West's will was certainly not restored by the feeble American responses to the 1979 seizure of U.S. diplomatic personnel in Tehran or by the unedifying spectacle of Western disarray over the question of whether to boycott Olympic Games hosted in a country in the process of attempting to crush armed resistance in Afghanistan. As James Schlesinger has noted, "A great power cannot retain influence when the conviction becomes widespread that it lacks the will to employ force to protect its interests."[12] Yet with the exception of France, since the OPEC oil embargo of 1973 no member of the North Atlantic Treaty Organization has shown a consistent willingness to employ force to protect interests outside the NATO area.

In the case of the United States, the issue is not confined to that of political will. Defeat in Vietnam and the disastrous military attempt in April 1980 to free U.S. hostages in Iran may have adversely influenced perceptions in some Third World countries of American ability to use force effectively. Indeed, these and other post–1945 military failures have led some American analysts to the conclusion that something far more fundamental is amiss with the U.S. military establishment than insufficient material resources or unwarranted political intrusion into operational planning.[13] In sum, even the reestablishment of substantial Western military capabilities in the non–Western world will be of little avail if those capabilities are not accompanied by a manifest willingness and ability to use them.

4. To assure their security NATO nations must strengthen peaceful ties with the Third World by means of increased trade, investment, economic aid, and where appropriate, security assistance. The steady decline over the past decade in the level of Western economic aid to, and investment in, the Third World, together with the persistence of formidable barriers to north–south trade may have contributed to the very instability in the non–Western world that the Soviet Union has exploited. With respect to security assistance, recent U.S. aversion to arms transfers outside the NATO area has served to deny weapons even to those who are prepared to resist direct Soviet invasion, such as Afghani forces resisting Soviet aggression. As noted by the Atlantic Council,

> from the point of view of many Third World governments seeking to maintain their independence, and feeling the need to turn to an outside nation for assistance, the failure of the Western nations to provide support in such circumstances seemed incomprehensible. To their eyes, it seems clear that if the presence of outside forces and weapons is not responded to by Western help, the balance of power in the situation will be determined by default. To them it seems clear that if Western military training is not available for their officer cadre or equipment for their troops, the Warsaw Pact will be ready to fill the

void. In such circumstances, the Third World nation may well consider the West an uncertain and unreliable ally, and turn away from the West toward accommodation with the East. Inaction, then, does not always equate with safety, as its proponents so frequently allege.[14]

What is needed is a coordinated program among the United States, its NATO allies, and Japan to provide selected Third World states with a level of economic and security assistance sufficient to promote internal political stability and a capacity for competent local self-defense. Such a program could draw heavily upon Japanese and European financial resources, and on North Atlantic weapons and military expertise.

OBSTACLES TO MEETING WESTERN SECURITY REQUIREMENTS OUTSIDE THE NATO AREA

It is far easier to define Western security requirements outside the NATO area than it is to fulfill them. Obstacles to meeting those requirements are formidable, challenging the very capacity of the West to provide a coordinated defense of its common interests in the non-Western world.

The problems confronting a reinvigoration of Western military power outside the treaty area are manifest in the rapid deployment force (RDF), America's chosen instrument for the defense of Western interests in the critical Persian Gulf region. The character and capabilities of the RDF merit considerable attention because: (1) with the exception of France, only the United States today deploys significant military forces outside Europe,[15] and (2) for a host of political, economic, and military reasons, the United States is the only member of the alliance capable of producing the additional forces and capabilities required for an expanded Western force presence and capacity to project force ashore outside the NATO area.

Deterrence of overt Soviet aggression in the Persian Gulf region and preservation of uninterrupted access to Persian Gulf oil remain the two principal rationales underlying the RDF, which was formed by the Carter administration in the wake of the Soviet invasion of Afghanistan in December 1979. An assessment of the utility of the RDF as a vehicle of U.S. military power must therefore be made in the context of the two pivotal conditions governing any U.S. attempt to mount a credible military defense of its vital economic interests in the Gulf: (1) the lack of secure military access ashore in the region and (2) unavoidable reliance on forces already committed to the defense of Europe, Northeast Asia, and other areas outside the Gulf region.

Secure military access ashore in the Persian Gulf is essential, certainly in contingencies entailing a prolonged land campaign. To get ashore intervention forces must have access to ports, airfields, and other reception facilities. To stay ashore they require continued access to nearby logistical support bases. Neither is available to the RDF in the Gulf region.

With the exception of the tiny atoll of Diego Garcia, some 2,500 miles from the Strait of Hormuz, the United States possesses no military bases in that vast area of the world stretching from Turkey to the Philippines. (In contrast are the large Soviet installations at Cam Ranh, Socotra, and Aden, and long-standing Soviet access to the Iraqi bases at Umm Qasr and Al Basrah.) Nor are prospects favorable for the establishment of a "Subic" naval facility or "Clark" air force base in the region. As confirmed in 1980 by then Under Secretary of Defense Robert Komer, "the countries [in the area] ... most emphatically do not want formal security arrangements with us."[16]

The political sensitivity of potential host nations to a permanent U.S. military presence on their own soil is certainly understandable and is manifest in their refusal to permit the peacetime stationing of any operationally significant U.S. forces on their territory. Such a presence "would validate the criticisms of radical Arabs about how the conservative [Persian Gulf] states are toadies of the imperialists," and thus "increase the chances of the internal turmoil that constitutes the main potential threat."[17] Moreover, many Persian Gulf states continue to regard U.S. support for Israel as a greater threat to the security of the Arab world than the prospect of an Afghanistan on the Arabian peninsula. Some even suspect the United States of coveting the peninsula's oil fields, a suspicion reflected in the following statement of Sheik Sabah al Ahmad al Sabah, Kuwait's foreign minister:

> Defend us against whom? Who's occupying us? We haven't asked anybody to defend us. Yet we find all these ships around asking for facilities. It's all a bit like a film with two directors—Russia and the U.S. How will the film end?

Perhaps with both big powers agreeing, 'O.K., these oil fields belong to us, and those to you. We'll divide up the region from here to there.' Is that how it will end?[18]

To its credit the U.S. Department of Defense appears acutely aware of the political barriers to establishing a permanent U.S. military presence ashore in the Gulf region and accordingly has pursued the alternative of gaining contingent rights of access to selected facilities in time of crisis. Agreements along these lines have been concluded with Kenya, Somalia, and Oman.

Yet simply having the promise of access to facilities on a contingency basis is no substitute for U.S. controlled and U.S. operated bases whose use is not subject to momentary political calculations of host governments. The same internal political considerations that deny the United States a permanent military presence ashore in the region could well be invoked to deny the United States access to facilities in the event of a crisis, irrespective of the agreements that have been negotiated.

The operational and logistical difficulties posed by lack of secure military access ashore in the Persian Gulf region are compounded by the lack of politically reliable and militarily effective U.S. client states in the region, whose assistance could be vital in a major contingency, particularly a prolonged one. The internal political fragility of potential U.S. friends and allies in the Gulf area is exacerbated by the questionable capabilities of their military establishments. The present Iraqi-Iranian war has done little to enhance the military reputation of the Arab world, and national military forces on the Arabian peninsula are negligible in size, questionable in quality, or both.[19] Enormous defense expenditures on the part of many Gulf states appear to have produced little in the way of technically competent, properly integrated, and well-led military forces. Even the large and experienced armies of Iraq, Iran, and Pakistan have been demoralized by defeat, revolution, or internal division along political or ethnic lines. In short, U.S. intervention forces could expect little effective support on the battlefield even from host nations requesting intervention.

The adverse consequences of local military ineffectiveness should not be underestimated, although U.S. intervention would of course benefit from it in contingencies involving aggression by a regional state. Although many commentators pronounced the Guam doctrine of the Nixon administration dead on arrival in the wake of the collapse of South Vietnam in 1975, the doctrine's fundamental premise remains as valid today as it did when promulgated in Guam in 1969: The sustained application of major U.S. military power in the Third World is not likely to succeed if unsupported by viable and competent local regimes capable of assuming a significant burden of the land battle. This is surely one of the principal geostrategic lessons of U.S. intervention in Indochina. To rush to the defense of any nation either unwilling or unable to defending itself is to rush into the abyss of another Vietnam. As then Secretary of Defense Harold Brown noted in testimony before the Senate Armed Services Committee in 1980, "the United States cannot defend . . . people in the [Gulf] region who are not willing to participate in their own defense. You need a significant political base and . . . effort by the people in the region."[20]

On what grounds, however, can the United States count on "a significant political base and . . . effort by the people in the region," particularly the kind of effort that would be required in the face of direct Soviet aggression or aggression by a Soviet client state? The availability of such support is ultimately a function of the political stability of the regime supplying it; its effectiveness is a product of the size and competence of the regime's military forces. In the Persian Gulf the West for decades enjoyed in the shah of Iran a powerful and seemingly stable local client committed to the defense of shared interests. Yet which potential Western client among the littoral states of the Persian Gulf and Indian Ocean today can be regarded as both politically stable and militarily competent? Somalia? Oman? Saudi Arabia? Kuwait? The United Arab Emirates? Pakistan? All of these states are governed by military regimes or semifeudal monarchies whose social and political fragility renders them vulnerable to internal overthrow by Soviet-sponsored leftist groups or the forces of religious fundamentalism now sweeping the house of Islam.

In short, in the Persian Gulf region the United States possesses none of the critical operational and logistical benefits that it enjoys in comparative abundance in Europe and in those other areas of the world such as Korea where large U.S. military forces are firmly ensconced ashore and can count on the support of powerful and reliable allies. As emphasized by Lieutenant General Paul X. Kelley, commander of the Rapid Deployment Force, the United States will have to "start from scratch" in the Persian Gulf against potential adversaries with large military forces already in place in the region or along its periphery.

> There are sizable U.S. forces in-place in Western Europe—with the exception of naval forces in the Indian Ocean, we have none in Southwest Asia.
>
> There are sizable amounts of prepositioned supplies and equipment in Western Europe for reinforcing units—we have none in Southwest Asia.
>
> There is an in-place command and control system in Western Europe—we have none in Southwest Asia.
>
> There is an extensive in-place logistics infrastructure in Western Europe—we have none in Southwest Asia.
>
> There are extensive host-nation support agreements between the U.S. and Western Europe countries—we have none in Southwest Asia.
>
> There is an alliance of military allies in Western Europe—there is no such alliance in Southwest Asia.[21]

Even were military access to the Persian Gulf not a problem, however, the commitment of any sizable U.S. force to combat in the region would automatically weaken the defense of no less critical U.S. interests elsewhere in the world. The decision to form the RDF from *existing* military units, almost all of which are already earmarked for NATO and the Far East, has served to widen what, even before Afghanistan, was a substantial gap between U.S. commitments abroad and capabilities to defend them. The decision makes it virtually impossible to deal effectively with a significant military challenge in more than one area at a time. As noted in 1980 by William Perry, then under secretary of defense for research and development, "the really troublesome problem we have is how do we accommodate [a] NATO buildup and the Persian Gulf buildup at the same time? That is the rub."[22] Army Chief of Staff Edward C. Meyer has stated flatly that present U.S. force levels are "not sufficient to repel a Soviet assault [in the Persian Gulf] without jeopardizing our NATO commitment."[23] Chief of Naval Operations Thomas B. Hayward has testified that the present "1½ ocean navy" of the United States cannot meet the "three-ocean commitment" imposed by the Carter doctrine.[24]

Although U.S. military planning since 1969 has called for capabilities adequate to wage simultaneously a major conflict in Europe and a lesser conflict elsewhere (the so-called 1½-war strategy), the simple truth is that U.S. force levels, now lower than at any point since the Korean War,[25] are not sufficient to meet the demands of more than one sizable conflict at a time. The ½-war in Vietnam was waged in no small part by U.S. forces earmarked for European contingencies. Similarly the two U.S. carrier battle groups now deployed in the Arabian Sea were drawn from NATO's southern flank and the Western Pacific.

Even the planned increases in the size of the U.S. Fleet from the present 440 vessels to 550-600 by the late 1980s is unlikely to be adequate. It is in any case highly doubtful whether the U.S. navy under the current all volunteer force is capable of manning a 600 ship fleet; existing shortfalls in skilled personnel are so severe that comparatively new vessels have been retired from active patrolling. With the exception of the marine corps, the other services are also suffering acute personnel shortages, especially among the highly skilled.

The strategic risk inherent in reliance on forces committed to both Persian Gulf and non-Gulf contingencies would be profound in circumstances involving a U.S.-Soviet confrontation. By virtue of interior lines of communication, larger forces, and greater proximity to both Europe and the Gulf, the Soviet Union could, by feinting in one area, divert rapidly deployable U.S. forces away from the true point of decision.

> The U.S. defense establishment currently lacks much latitude to cope with sizable contingencies [in the Persian Gulf]. The Soviet side, whose large forces afford more flexibility, could sponsor several widely-separated hot spots at the same time, with assistance from allies and friends.
>
> Possible application of U.S. military sinew to ensure petroleum imports from the Persian Gulf should be viewed in that perspective.
>
> Our active status strategic reserves are too few to fight even a modest war in the Middle East without accepting calculated risks that uncover crucial interests elsewhere. Even "best case" forces would probably prove insufficient against the Soviets, whose abilities to project offensive power beyond their frontiers have improved impressively in recent years.[26]

This is not to suggest that present U.S. forces could be no better postured than they are today. On the contrary, the readiness, structure, weaponry, and logistical sustainability of many units are not optimized for many conceivable contingencies in the Gulf region. More fundamental is the comparatively small amount of real combat power generated by the present structure of U.S. military forces. The product in large part of low ratios of combat to support and reliance on individual (versus unit) replacement policies characteristic of a continuing orientation toward protracted conflict and a firepower/

attrition style of warfare, the problem is nowhere more visible than in the U.S. army. In contrast to the Soviet army, for example, which fields a total of 173 divisions from a manpower base of 1.8 million active duty personnel, the U.S. army musters but twenty-four active and national guard divisions from a base of 775,000 active duty personnel. (Although the personnel strength of Soviet divisions is approximately 70 percent of the U.S. counterparts, in terms of tanks and other major items of equipment they are about equal.) Finally, the present acute shortage in strategic lift capabilities, which cannot be overcome for at least a half-decade, constrains U.S. ability to move military forces into the Gulf in a timely fashion.

Thus existing U.S. forces are simply insufficient in size to meet the requirements of defending the Gulf without reducing the capacity of the United States to meet its commitments elsewhere. In the absence of (1) larger forces, (2) forces fundamentally restructured to maximize combat power, or (3) expanded European forces dedicated to defense of the Persian Gulf or a stronger defense of Europe, the price of effective deterrence in the Persian Gulf is a degradation of deterrence in Europe and Northeast Asia.

It might have been assumed that the problems of access and insufficient force would have impelled the Pentagon toward both larger force levels and the creation of an instrument of intervention based primarily on sea power and supplemented by robust amphibious assault and other forcible-entry capabilities. An RDF of this kind would be comparatively free of dependence on the presumed political goodwill of unstable, potential host governments in the Persian Gulf region and able, if necessary, to gain access ashore without invitation.

Such unfortunately has not been the case. In terms both of forces earmarked for the rapid deployment mission and of the means chosen to enhance their ability to deploy rapidly to the Gulf, the present RDF's utility is questionable in contingencies not involving access ashore, willingly granted, before hostilities.

As shown in Table 6-1 most of the military units now assigned to the RDF are not only already committed to NATO but also dependent for entry ashore and subsequent sustainability on friendly ports, airfields, and logistical facilities. Only the carrier battle groups and the marine amphibious force are deployed or deployable afloat and logistically supportable from the sea; only the marines and the army's 82nd airborne division possess the ability to enter territory controlled by hostile forces. The remaining units earmarked for the

Table 6-1. Tentative Rapid Deployment Joint Task Force Composition, 1981.

Unit	NATO Earmarked?	Land-Dependent?[a]
Ground Forces		
(Army)		
18th airborne corps HQ	no	yes
82nd airborne division	yes	yes
101st air assault division	yes	yes
9th infantry division	yes	yes
24th mechanized division	yes	yes
194th armored brigade	no	yes
6th cavalry (air combat) brigade	yes	yes
two ranger infantry battalions	yes	yes
(Marine Corps)		
one marine amphibious force	?[b]	no
U.S. Air Force Units		
one air force HQ	?	
twelve tactical fighter squadrons	yes	yes
two tactical reconnaissance squadrons	yes	yes
two tactical airlift wings	no	yes
U.S. Navy Forces		
three carrier battle groups	some[c]	no
one surface action group	some[c]	no
five aerial patrol squadrons	some[c]	yes

a. Dependent for commitment ashore on access to secure ports and/or airfields, *or* dependent for subsequent operations ashore on a shore-based logistical infrastructure.

b. Marine amphibious forces are considered available for contingencies worldwide.

c. A significant proportion of U.S. naval vessels currently deployed in the Indian Ocean belong to the U.S. 6th fleet in the Mediterranean. Maintenance of a naval presence in the Indian Ocean at the expense of NATO-oriented naval forces is likely to continue throughout the decade.

RDF, which organizationally consists of little more than a new headquarters charged with identifying, training, and planning the employment of existing forces suitable for Persian Gulf contingencies, are completely land-dependent.

The presumption that the RDF will enjoy uncontested entry ashore in the Gulf is further apparent in the strategic mobility programs associated with the force and in the absence of any proposed increases in amphibious assault capabilities. To speed the deployment of RDF forces from the United States to the Gulf in a crisis, the Pentagon has requested funding authority to develop and produce a new strategic air transport (the CX); a flotilla of fast sea transports; and a total of twelve specialized logistics ships, known as maritime pre-positioning ships (MPSs), aboard which will be stored equipment and 30 days' worth of combat consumables for a complete marine division. The MPS vessels, which will be maintained on station in the Indian Ocean, represent partial compensation for the inability to pre-position equipment and supplies ashore.

None of the proposed ships and aircraft, however, possess any defensive capabilities; all will be unarmed, and the ships will be manned by civilian crews. Hence their utility, like that of the bulk of the forces they will be carrying, is dependent on friendly reception ashore.

This is not to suggest that expanded sea power alone could ever provide a credible deterrent to the full spectrum of military threats confronting the West in the Persian Gulf. A sea-based RDF would have limited utility in contingencies demanding sustained combat inland, beyond the reach of amphibious assault forces and carrier-based air power. Any sizable direct conflict with Soviet forces in the region certainly would demand large, heavy U.S. ground forces and massive quantities of U.S. land-based tactical aviation. Prosecution of sustained inland combat, however, would be contingent upon secure coastal military lodgements, which, in lieu of access ashore before hostilities, could be gained only by the ability to project power ashore. Moreover, unlike land-based tactical air forces withheld in the United States for rapid deployment to the Persian Gulf region, a sea-based RDF would have the advantage of already being there. In short, as long as U.S. military forces are denied politically secure access ashore in the Gulf region in peacetime, there appears to be no alternative to heavy reliance on sea power, at least as the critical cutting edge of an intervention force.

NEW TASKS FOR THE ALLIANCE

Given the boundaries of the NATO treaty area and the predominance of U.S. forces among Western military power deployed outside the area, any effective Western defense of its interests in the Third World must be predicated on a new division of military labor within the alliance. To be specific, if the United States is to assume the main burden of meeting Western security requirements outside the North Atlantic area, then NATO Europe must be prepared to assume a large proportional share of the conventional defense of Europe.

Transatlantic policy differences over relations with the Soviet Union cannot be allowed either to obscure the fundamental threat to Western security posed by Soviet actions outside Europe or to stifle attempts to respond to those threats in a coordinated fashion. The West should continue to explore avenues of cooperation with the East; but it also must be prepared to compete with Moscow in areas where competition, military or other, is discernibly the only means of preserving Western security interests. It is again worth restating that both Europe and Japan are far more dependent than the United States on secure access to the oil and raw materials of the non-Western world. It is further worth recalling that the United States continues to bear, as it always has, the primary burden of strategic and theater nuclear deterrence in both Europe and Northeast Asia as well as the main burden of Japan's conventional defense. Finally, it is important to recognize that traditional isolationist sentiment is only dormant—not dead—in the United States, and that neither the Congress nor the American taxpayer can be expected to tolerate indefinitely a situation in which a U.S. move to provide an adequate defense of common Western interests outside the NATO treaty area is unaccompanied by a compensatory redistribution of burden-sharing within the alliance. In this regard, the failure of European members of NATO to meet pledged real annual increases in defense spending of 3 percent certainly would not be well received by a U.S. Congress that is being asked by the Reagan administration to authorize annual real increases in excess of 7 percent in U.S. defense spending, much of it to protect NATO interests, over the next several years.

Yet pledged real increases of 3 percent may not be sufficient to finance even those force improvements contemplated in the pre-Afghanistan long-term defense plan, to say nothing of additional forces and capabilities required to "take up the slack" in Europe caused by the reallocation of U.S. forces originally designated for NATO to missions outside Europe. The ravages of inflation, the relentless growth of Western military personnel costs, and the steady real increases in Soviet defense outlays, on the order of 4 to 5 percent per year, argue strongly for NATO increases considerably higher than 3 percent if NATO is to meet the burgeoning Soviet military challenge in Europe alone. It is generally recognized that the figure of 3 percent is an arbitrary one, chosen largely for political reasons and bearing little if any analytical relationship to the anticipated costs of financing the long-term defense plan.

A new division of military labor within NATO does not mean that the United States can or should go it alone in the Persian Gulf and elsewhere in the Third World. The fact that NATO as an institution is ill-suited to address the security challenges facing its members outside the treaty area does not preclude cooperation outside the treaty area on an ad hoc basis among individual members. Such cooperation is both desirable and feasible, as demonstrated by Franco-American cooperation in Shaba in 1978, and continuing U.S.-British cooperation with respect to Diego Garcia.

Many European members of NATO maintain, as does the United States, strategically mobile forces capable of rapid deployment outside Europe. France, Belgium, Germany, Italy, and the United Kingdom possess sizable airborne formations along with modest airlift capabilities. Special commando units also may be found in the British, Belgian, and other NATO European armies. Several NATO members possess significant numbers of maritime surveillance/patrol aircraft as well as oceangoing minesweepers and mine warfare vessels capable of supplementing the limited inventories available to the United States. Additionally, the major European members of NATO (and Japan) have substantial sealift resources that could prove invaluable during a deployment crisis in the Gulf region. Nor should European capabilities to project power ashore be overlooked: the French and British navies contain a total of six aircraft carriers[27] and two battalions of marines.

Indirect potential European contributions to a defense of Western interests in the Gulf and elsewhere outside the treaty area include provision of airfield and port facilities (both in Europe and the Middle East) and financial support to RDF-related military construction projects in the Indian Ocean.[28] RDF force planners also would benefit immensely from France and Britain's greater knowledge of and experience in certain areas of the Third World, notably Africa and the Arabian peninsula. For example, British knowledge of the political, social, cultural, and military climate of Oman vastly exceeds that of the United States; British military personnel still comprise one-half of the officer corps of the Omani army, and the Omani government continues to retain a number of high-ranking British advisors. The French enjoy a similar status in several French-speaking African states.

In sum, NATO Europe has an important contribution to make to the defense of Western interests outside the NATO area, although such a contribution must be made, as it has been in the past, outside the institutional framework of the NATO alliance. While the question of burden-sharing with respect to the defense of Europe inside the treaty area will remain on the agenda of NATO as a collective security organization, the question of defending Europe outside the treaty area must of necessity be addressed by an informal coalition of willing members. To be effective, such a coalition demands both a continual exchange of information and periodic if informal consultation.

The United States of course must also do its part; like NATO Europe it too must be prepared to make unpleasant budgetary and other sacrifices. As the principal Western military actor outside the NATO area, it is incumbent upon the United States to devise an effective and genuinely credible military instrument for the defense of Western interests in the Third World. A rapid deployment force whose employment is excessively dependent on friendly invitation and would automatically degrade the defense of Europe simply will not suffice. Nor will it be possible in the long run for the United States to meet its burgeoning extra-NATO military responsibilities within the framework of the all volunteer force.

NOTES TO CHAPTER 6

1. James R. Schlesinger, "The Geopolitics of Energy," *The Washington Quarterly*, 2 (No. 3, Summer 1979): 7.
2. See, for example, Robert J. Hanks, *The Unnoticed Challenge: Soviet Maritime Strategy and the Global Choke Points* (Cambridge, Mass.: Institute for Foreign Policy Analysis, 1980).
3. Chris L. Jeffries, "NATO and Oil," *Air University Review*, 31 (No. 2, January-February 1980): 44.
4. Albert Wohlstetter, "Half Wars and Half Policies in the Persian Gulf," in W. Scott Thompson, ed., *National Security in the 1980s: From Weakness to Strength* (San Francisco: Institute for Contemporary Studies, 1980), pp. 145-146.

5. Richard Burt, "Washington and the Atlantic Alliance: The Hidden Crisis," in Thompson, ed., *National Security in the 1980s*, pp. 110-111.
6. C. Kenneth Allard, "A Clear and Present Danger: Soviet Airborne Forces in the 1980s," a paper presented before the Conference on Projection of Power: Perspectives, Perceptions, and Logistics, Ninth Annual Conference of the International Security Studies Program, The Fletcher School of Law and Diplomacy, Tufts University, Boston. April 23-25, 1980.
7. Jeffrey Record, *The Rapid Deployment Force and U.S. Military Intervention in the Persian Gulf* (Cambridge, Mass.: Institute for Foreign Policy Analysis, 1981), p. 14.
8. *Defense in the 1980s, Statement on the Defense Estimates, 1980, Volume 1* (London: Her Majesty's Stationery Office, 1980), pp. 40-41.
9. Amphibious ships are essential to the projection of ground forces from ship to shore against a hostile beachhead. The present U.S. level of sixty-four vessels is sufficient to lift only one of the U.S. marine corps' three divisions. However, because the ships are widely dispersed throughout the U.S. navy's various commands, no more than a battalion landing team's worth of shipping is usually available at any given location. To assemble the shipping required for a division-size amphibious assault would require 30-45 days.
10. "Our existing mobility forces cannot meet the deployment objectives we have set... for NATO or for some non-NATO contingencies." Harold Brown, *Department of Defense Annual Report, Fiscal Year 1981* (Washington, D.C.: U.S. Government Printing Office, 1980), p. 208. The magnitude of current shortfalls in U.S. strategic lift capabilities is evident in U.S. plans to double the number of U.S. divisions in Europe within 10 days of a decision to mobilize for a NATO contingency, and in plans to procure an additional 150-200 strategic lift aircraft, add eighteen tankers and dry cargo ships to the ready reserve fleet, and construct twelve specialized maritime pre-positioning ships for non-NATO contingencies. Enhancement of rapid reinforcement capabilities is also a major objective of NATO's long-term defense plan.
11. *After Afghanistan — The Long Haul, Safeguarding Security and Independence in the Third World* (Washington, D.C.: The Atlantic Council of the United States, 1980), p. 13.
12. James R. Schlesinger, "Some Lessons of Iran," *New York Times* (May 5, 1980): A32.
13. See Jeffrey Record, "Is Our Military Incompetent?" *Newsweek*, (December 22, 1980), and "The Fortunes of War," *Harpers* (April 1980); Edward N. Luttwak, "The American Style of Warfare and the Military Balance," *Survival* (March/April 1979) and "The Decline of American Military Leadership," *Parameters* (December 1980); Richard A. Gabriel and Paul L. Savage, *Crisis in Command, Mismanagement in the Army* (New York: Hill

and Wang, 1978); Cincinnatus (pseudonym), *Self-Destruction: The Disintegration and Decay of the United States Army during the Vietnam Era* (New York: Norton, 1981); and Steven L. Canby, "General Purpose Forces," *International Security Review* (Fall 1980).
14. *After Afghanistan*, p. 41.
15. France deploys sizable military forces outside the NATO area, including a large standing naval presence in the Indian Ocean and over 8,000 ground troops in French-speaking Africa, where she has bilateral defense agreements with Cameroun, Benin, the Ivory Coast, Niger, Mauritania, Senegal, Togo, the Central African Empire, the Congo, Gabon, and Djibouti. Major French ground deployments in Africa include 4,150 troops in Djibouti and 1,170 in Senegal. French willingness to use force on behalf of threatened interests in Africa was demonstrated in Chad during the 1970s, when French troops were employed to support the suppression of Libyan-sponsored guerrilla activity; in 1978 in Zaire, when a French parachute battalion intervened at Kolwezi to rescue European survivors of an attempted Angolan incursion; and in 1980, when French naval and air units were deployed in Tunisian waters in response to a Libyan-backed incursion into that country.
16. *Department of Defense Authorization for Appropriations for Fiscal Year 1981, Hearings before the Committee on Armed Services*, U.S. Senate, 96th Congress, 2nd Session (1980), part 1, p. 445 (hereinafter cited as *Hearings*).
17. Richard K. Betts, *Surprise and Defense: The Lesson of Sudden Attacks for U.S. Military Planning*, a draft manuscript scheduled for publication by the Brookings Institution, pp. ix–14.
18. "Preserving the Oil Flow," *Time* (September 22, 1980): 29.
19. See Abdul Kasim Mansur (pseudonym), "The Military Balance in the Persian Gulf: Who Will Guard the Gulf States from Their Guardians?" *Armed Forces Journal* (November 1980).
20. *Hearings*, part 1, p. 35.
21. Statement by Lieutenant General P. X. Kelley on rapid deployment force programs, before the U.S. Senate Armed Services Subcommittee on Sea Power and Force Projection, March 9, 1981, p. 5.
22. *Hearings*, part 6, pp. 3, 275.
23. Ibid., part 2, p. 745.
24. Ibid., p. 785.
25. A benchmark widely used for comparison with the present is 1964, the last year preceding the U.S. buildup in Vietnam. For example, in 1964 personnel assigned to the active armed forces totaled 2.685 million compared to 2.059 million in fiscal year 1981; the active U.S. fleet contained 803 vessels compared to the present 440; the U.S. air force fielded a total of 439 active squadrons of all types compared to 253 today; and strategic

sealift vessels numbered 100 compared to the present 48. In 1964 the United States allocated 8.4 percent of its gross national product to defense compared to less than 5 percent in the fiscal years 1977–1980.

26. John M. Collins et al., *Petroleum Imports from the Persian Gulf: Use of U.S. Armed Force to Ensure Supplies* (Washington, D.C.: Library of Congress Congressional Research Service, 1980), p. 16.
27. All are small compared to U.S. carriers, and Britain's three carriers are designed primarily for antisubmarine warfare.
28. According to sources in the U.S. Department of Defense, identifiable military construction costs associated with planned U.S. improvements in facilities to which the United States has been granted access in Oman, Kenya, and Somalia will total $374 million for the period of fiscal years 1981–1983. Military construction projects also are planned in Egypt.

7 CONCEPTS AND CAPABILITIES

Russell E. Dougherty

THE FUNDAMENTALS OF NATO'S DETERRENT CONCEPT

Given the strategic concept of the NATO alliance and the military capabilities nations make available to implement that concept, the resultant credibility of the NATO deterrent would seem to be simple, calculable, and precise. In truth, too many of us have always succumbed to the temptation of just such a simplistic calculation in assessing the credibility of the NATO deterrent. Overlooked is the fact that NATO's strategic concept is basically political; and the capabilities that result must in the final analysis be constrained in type and quantity by what is politically acceptable to a diverse group of sovereign allies. Strong, cohesive political leadership throughout the alliance is essential to a proper balance of the will and the capabilities NATO evidences, and such stability or consistency has not yet been evidenced by NATO's members. The continued credibility of our NATO deterrent is in question.

Assessing only the concept and the capabilities of NATO is not an adequate basis for judging its overall credibility as an alliance for two reasons, the first of which is extremely important, but the second of which is critical.

What is important is that assessing the credibility of a deterrent posture requires a comprehensive and accurate understanding of who is to be deterred, from what and where. Although NATO has succeeded in deterring the USSR from direct military attacks on the sovereign territory of its members, those nations must recognize that the Soviets have succeeded in eroding our unity and cohesion in peripheral areas, that our resources are threatened, and that these limited constraints will not serve us well against the increasing, coherent Soviet threats of future years. The United States have failed to comprehend the totality of the political strategy of the Soviets toward the West and have underestimated the warfare doctrines they employ in consonance with that total threat. Americans have failed to grasp the full significance of the "correlation of forces" concept of the USSR and the dynamic and relentless growth in that nation's overall capability for projecting its total strength and for total war, nuclear and conventional, if war should be "forced" upon them as a result of the success of their political aggression. We are not alone in this failure to understand our enemy; it pervades the alliance.

What is critical is that the democratic composition of the nations of the NATO alliance demands that the concepts NATO espouses and the capabilities employed to implement the concepts be understood and supported widely by public opinion throughout the alliance. Societal acceptance is critical to the success of the alliance. Obtaining and retaining widespread support for military concepts and capabilities is imperative. Without such popular support and without a continuing pattern of cohesion and concern among the key opinionmakers throughout the alliance, NATO cannot expect to be credible as a deterrent to aggression or war, particularly in an era of complex and dynamic changes and challenges all over the globe. Critical to NATO's deterrent posture is the demonstrated will of the member nations to concert all aspects of their political strength in support of NATO's strategic concepts; the will to maintain, train, and equip relevant armed forces in peacetime and to permit their optimum disposition and deployment. Unfortunately throughout NATO there is a widespread and debilitating public detachment from the requirements for future deterrence. This public detachment is at once political, psychological, economic, and military. Continued apathy could doom the effectiveness of NATO's deterrent posture. It is incumbent upon the leadership of NATO to elicit more support for NATO's deterrent posture and assure that both the will and the flesh are strong enough to match Soviet political threat.

EVALUATION OF NATO'S STRATEGIC CONCEPT

A backward look at NATO's conceptual history is instructive; a forward look is imperative. The preceding chapters of this assessment of the NATO nations' prospects for future security vividly illuminate the proposition that we must do more and do it better if we are to continue to have the political freedoms our political philosophies and our national interests require. The question remains what should be done *differently* to improve NATO strategy.

In the beginning (1949-1954) NATO grasped for what seemed to be reasonably available and necessary to stem the momentum of post-World War II Soviet encroachment into Western Europe. Politically, cohesion and popular support were sought through emphasis on what General Dwight Eisenhower called "the impulse of fear" and through recognition that the Soviet's westward thrust could not be stopped by any one ally. Thus, NATO's members first banded together to do what no one of the nations could do alone. It was deemed imperative to stop the Soviets at the border of NATO's territory. A politically inspired concept of "forward defense" was adopted that envisioned a composite *causus belli* if any part of NATO's territory or NATO's forces were attacked. Militarily, a very large conventional force requirement was then established for the emerging NATO military commands (the seventy, plus, divisions of the 1952 Lisbon force goals) that would be designed and positioned to provide a countering defensive barrier along the entire continental periphery of NATO; with consequential reserves and supporting forces, in keeping with sound military principles. No formal strategic conceptual document was adopted during these early years.

The political disconnection between the initial, informal strategic concept and the willingness of nations to provide the military capabilities deemed necessary to support that concept was almost immediate. The situation was attenuated, however, by the Korean conflict in the early 1950s, which resulted in some U.S. reinforcements being sent to Western Europe and the rapid development of increased long-range nuclear forces (B-29s/B-50s) by the U.S. strategic air command (SAC), many of which were positioned forward in the United Kingdom to deter any Soviet moves into Western Europe during the Korean War years. This long-range "retardation force" of U.S. bombers caused many NATO policymakers to recognize the possibility of compensating for the evident unwillingness of the struggling economies and exhausted peoples of NATO to provide large, standing conventional forces by substituting the threat of external nuclear forces (primarily SAC's long-range aircraft) to prevent or blunt a major Soviet attack on NATO. This seemingly low-cost solution (not personnel-intensive nor demanding an extensive European involvement) had widespread appeal and rapid acceptance as a defense philosophy for NATO. In December 1952 the North Atlantic council, which oversees NATO, approved military committee document 14/1 (MC-14/1); the first formal articulation of NATO's strategic concept of large-scale retaliation as a basis for its defense.

But time was to pass before that concept was accepted. In 1953 the North Atlantic council noted that the United States was seeking legislative authority to share nuclear information with the allies, a matter prohibited by the original Atomic Energy Act.

The confirmation that NATO needed in order to revise its political thinking about the need for massive conventional armies, navies, and air forces, came with Secretary of State John Foster Dulles' 1954 articulation of the massive retaliation strategy to be pursued by the United States (the promise of the nuclear sword, later to be moderated in NATO parlance to the "strategic umbrella"). With this confirmation of a defense concept by the United States, the logic for reliance on clear U.S. nuclear superiority for deterrence of the inadequately equipped Soviet forces was established.

At the end of 1954 NATO authorized military planning to take account of "modern developments in weapons and techniques," a euphemism well understood to mean nuclear weapons. But at the same time it was found necessary to reassure the European citizenry that the military authorities of the alliance were not being given the authority to put plans into action.

In the meantime the relatively small in-place conventional forces of the alliance were postured in a thin forward defense near the political borders to establish the fact of aggression and encroachment. Air defense forces were developed and placed under SACEUR's operational control in 1955 by authority of MC-54/1, "Integration of Air Defense in NATO."

In 1956 the NATO political directive to the military authorities called for conventional resistance to "infiltrations, incursions and local hostile actions" but authorized planning on nuclear weapons in other circumstances. This somewhat ambiguous doctrine was fleshed out in the revised strategic concepts of MC-14/2, December 1957, and MC-48/2, dealing with a strategic concept and doctrine that favored early use of nuclear weapons. Most important, governments on both sides of the Atlantic did not plan for or invest more heavily in conventional forces. Nuclear weaponry was given further emphasis in 1957 when the NATO nuclear stockpile was established, and the emplacement of intermediate-range ballistic missiles in Europe was agreed upon. Thus the external nuclear force of the U.S. and a small but evolving multinational force of nuclear-capable fighter–bombers, armed with nuclear weapons maintained by U.S. custodians, became the deterrent threat to any major military aggression.

There was much room for interpretation of NATO doctrine as to what level of conventional force was needed. Many European nations in general interpreted it to sanction something akin to a "tripwire." Secretary Dulles strongly opposed that interpretation and, to the surprise of some who saw him as the father of the doctrine of massive retaliation and hence the disregard of conventional forces, pushed for a better balance between conventional and nuclear force, as did every SACEUR. These pressures were reflected in NATO communiques of subsequent years; for example, in 1960: "There must be a proper balance in the forces of the Alliance of nuclear and conventional strength to provide the required flexibility. The Ministers, in light of the Annual Review, took note of the progress which had been made, and expressed their determination to continue their efforts to improve the deterrent and defensive strength of the Alliance." But the measures undertaken by the member nations to build up conventional strength were in fact inadequate.

Throughout the 1950s and 1960s, the sound strategic thinking of Generals Gruenther, Norstad, and Lemnitzer (successive supreme allied commanders for Europe) caused them to caution nations repeatedly against thinking that adequate deterrence could be achieved without a real warfare capability. They pressed for realistic, dynamic modifications in conventional equipment, substantial deployments, and public understanding of the limitations of both the nuclear and conventional forces. They correctly interpreted the need for continuing modifications and increased flexibility in our strategic concepts and force characteristics to counter the changes they saw emerging in Soviet equipment and dispositions, while preserving the political imperative of forward defense. The ebb and flow of the equipment and force posture of the alliance reflected the overriding and diverse economic and political concerns of nations. Fortunately the relative Soviet force improvements, though substantial, did not reach dramatic proportions until 1967-68; thus NATO was able to weather internal discords concerning MC-14/2 as well as the recurring Berlin crises happening a few miles from the politically drawn West German border of the alliance.

The military necessity for adopting a more flexible (less dogmatic) strategic doctrine for determining the character and weight of NATO's response to aggression and attack on its territory gained powerful emphasis in the political environment of the Kennedy administration. We tried to design a strategy to make sense—to provide for defense forces, covering forces, maneuvering forces, reinforcing forces and both conventional and/or nuclear striking forces. Support for flexibility in our response came from very different motives and widely diverse understanding of the "why" of flexibility. Tailoring the ability to respond based on the extent of provocation or aggression gained acceptance by some and was seen as an invitation to aggression by others. Notwithstanding these significant differences in motivation, the idea of flexibility in strategic doctrine gained strong political impetus and support. Regrettably, the evolution of military capability, command and staff structure, and the extent of the supporting environment did not keep pace with the changes in NATO's strategic thought. This was so almost to the point of complete disconnection in 1967, and it is so to a significant degree today.

The government of France objected vigorously to any change in NATO's MC-14/2 strategic concept. This objection was voiced consistently in every forum, every military exercise, through every channel available. The tide of change was too strong politically and too logical militarily for France to contain. In the mid-1960s the future credibility of NATO's automatic nuclear sword was being questioned; the inadequacy of NATO's conventional shield forces was starkly apparent. Soviet force improvements in nuclear and conventional forces demanded a change, and, practically, the national strategic planning in the United States was undergoing an evolution toward greater flexibility in concept and in characteristics of its nuclear forces.

The process of change in NATO's massive nuclear retaliatory concept (MC-14/2) was underway at international action levels, spurred on by frequent policy level interventions. A major attempt to align national force planning and NATO's military requirements was initiated in 1964; however, French objections to any modification of MC-14/2 served to bar consensus within the NATO Atlantic Council on a coherent five-year force planning or on the revision of NATO's strategic concept. The 1966 decision of the government of France to withdraw its forces from the integrated force structure and international military staffs of the alliance in 1967 had been forecast by General Charles DeGaulle almost a decade earlier in a speech at the Ecole Militaire in October 1958: "This subordination known as integration must end." The French elected to assume unfettered national sovereignty over the NATO bases and logistic environment within France, to request NATO's political and military headquarters to relocate from France, and to disassociate France from all NATO's integrated military planning (except air defense, where close cooperation was obviously in France's self-interest). Thus they made possible the adoption by "the Fourteen" (NATO less France) of a more flexible strategy, without fear of a French "veto." (In practice, NATO has never adopted a formal position or decision on matters affecting the alliance over the overt objection of one of its members. In the matter of the revision to NATO's strategic concept, it was evident that France's position was not one of tacit acceptance; it was one of strong, overt objection.)

Military committee document 14/3 (MC-14/3) was adopted by the newly created group of fourteen, sitting without France as the NATO Defense Planning Committee (DPC), meeting in the NATO headquarters in Brussels in December 1967. (The DPC was created to handle the military aspects of NATO's integrated forces—land, sea and air. France's withdrawal from such matters did not constitute withdrawal from the alliance. France still sits as a member of the North Atlantic Council).

MC-14/3 presented the so-called flexible response strategy of NATO's international, integrated military forces, which remains NATO's strategic concept. It continues to reflect the fundamental political consensus of the alliance less France. Whether this strategic concept will continue to serve well is a question that deserves examination; it is not as inflexible and restrictive as the earlier policy, permits discretion in NATO's military response to armed attack against a member nation, and is more sensitive to changed circumstances, NATO's or the adversary's, than the policy it replaced. To paraphrase General Bernard Rogers, SACEUR, speaking to the Supreme Headquarters Allied Powers Europe (SHAPE) *Anciens* in Washington in November 1980: "We've got a strategy—it remains to be seen whether we in NATO have the will and the determination to provide the capabilities to insure that it serves us well throughout the years ahead."

THE CURRENT NATO STRATEGIC CONCEPT

An accurate unclassified description of NATO's strategic concept must start with the observation that MC-14/3 articulates only a defensive concept; it does not envision any unprovoked, offensive use of NATO's political or military capabilities for aggression. It establishes the primary objective of NATO's collective strengths, including its military forces and plans for their employment, as *deterrence*. Deterrence of major attack, of course; but also deterrence of coercion; intimidation, and political pressures made intolerable through military threats.

The concept very properly recognizes the sine qua non of deterrence as the product of the capabilities and the collective political will of NATO's nations to respond by all means available to any external threat to the territory and sovereignty of its member states, which are established as the vital interests of the alliance. Also the concept articulates the political imperative of its member states, particularly the European members, to ensure that the military strategies adopted by the major NATO commanders will be premised on a "forward defense." Though a forward defense in Western Europe may be politically demanding, there are in fact few areas in NATO Europe where territorial room for maneuvers are meaningful, militarily, in the face of modern weaponry. Except for the terrain of France and unaligned Spain, there is little depth in the integrated geography of NATO Europe to permit secure reception facilities, redoubts, and extensive maneuver for modern, mobile military forces.

It is evident that an absolute walllike forward defense is not possible along the peripheral borders of the alliance. It is accepted, however, that NATO's defensive concept does not envision trading any substantial portion of any member's territory for time or tactical advantage.

As General Lauris Norstad observed: "we cannot omit or skimp at any point along our Eastern frontiers the defense strength which the Soviet threat dictates. For if our line is not defended throughout, the enemy might trump up a pretext for crossing it. We would then face not only an accomplished fact, but also a dilemma. If we did not take immediate action, we would fail to meet the commitments of the Alliance; if we did take it, we would start a war."[1] To the extent possible, main defenses must be planned as close to political borders as possible and no hostile lodgements will be condoned or permitted without immediate, defensive response.

The duration of conflict is not predetermined in NATO's strategic concept, though it is envisioned that, if deterrence fails and forward defenses are inadequate to prevent incursions, sufficient forces will be employed to contain the conflict at the lowest level possible consistent with the objectives of dislodging any occupying enemy and restoring the territorial integrity of the alliance. In this regard the option for escalation is specifically preserved, and a conceptual linkage is established among the conventional forces of the alliance, the theater nuclear forces, and external strategic nuclear forces.

In the employment of its forces NATO's strategic concept envisions a collective application of force through the strategy of flexible response, often the subject of debate, disagreement, and misunderstanding. Fortunately no divisiveness is apparent in the consistent pattern followed by SACEURs Lemnitzer, Goodpaster, Haig, and Rogers in their planning for implementation of this key aspect of the NATO concept. Direct defenses and counterattacks to aggression are planned at a level appropriate to the scale of attack; but with NATO plans for deliberately escalating the weight of attack, broadening the area and scope of conflict, or using nuclear weapons in engagements ranging from a limited battle to general nuclear war. In sum, the concept envisions NATO forces fighting effectively at any level of conflict, or escalation to another level of conflict (through NATO's own volition or in response to escalation by the enemy), until the conflict can be terminated and the aggressor repulsed on terms that preserve the vital collective interests of the alliance. General Lemnitzer cautioned his commanders in 1968 not to confuse the NATO concept of flexible response with some form of military tit-for-tat, or gradualism.

Thus NATO's 1967 strategic concept for deterrence and defense of NATO territory is one of: (1) collective NATO action, (2) forward defense, (3) flexible response, (4) preparation for a defense of indefinite duration, and (5) deliberate escalation where it is in NATO's interest. The concept is, as General Bernard Rogers has called it, "as valid today as when it was first agreed within the Alliance." In support of General Rogers' view, it must be reported that few if any senior officers with NATO experience have found serious flaws in the basic conceptual framework of NATO's strategy. Nevertheless, many military and political officials are extremely concerned with the precise geographic constraints that have evolved prescribing NATO's nations' collective obligations to each other.[2] There is widespread concern that the nature of the Soviet threat has made such limited area interpretation of NATO's collective defense obligations passé. It is to be hoped that NATO's collective will to respond by all appropriate means would transcend the precise constraints of Article 6 of the North Atlantic Treaty (quoted in note 2) and be expanded to encompass the defense of vital interests of the alliance nations wherever located. The political problems this presents are recognized; nonetheless, the Soviet political thrusts and the burgeoning Soviet capabilities for global interference and aggression are unfettered by any such geographical constraints, and unless appropriate action is taken, we may find the NATO nations deterred from individual responses, let alone a collective response in matters of vital interests arising outside the prescribed NATO area.

CONCERNS ABOUT NATO'S STRATEGY AND ITS CAPABILITIES

Most of the frequently voiced questions and objections to NATO strategy do not stand up to close scrutiny; usually such objections are premised on or have resulted from misinterpretation of the great flexibility in the strategic concept, improper or inadequate capabilities to implement adequate defenses, or inadequate supporting arrangements. Consider the following:

- Is the strategy still viable? *Answer*: Only if the alliance is politically and militarily viable and only if the leadership and important political segments of NATO's member nations evidence their willingness to provide and maintain the capabilities to support the strategy and keep them relevant to the changed circumstances of a dynamic threat. This requires enlightened political decisions within and among the allies and long-term planning and consistent action. To the extent that the NATO nations postpone or procrastinate on the provision of critical capabilities and support can erosion in the credibility of NATO as a deterrent be expected. Such erosion cannot properly be attributed to an inadequacy in the strategy of the alliance; rather, it results from failure to provide the essential guidance and provide appropriate capabilities to implement the strategy.

• Is it a realistic tenet of our strategy to consider a limited—or any—use of nuclear weapons in defense of NATO? *Answer*: Yes. There is no real alternative to that tenet. Given the comprehensive nuclear capabilities of the USSR it is unrealistic *not* to provide modern, relevant nuclear capabilities and plans for their use. It is also unrealistic to expect NATO's deterrent force to be operative in a situation of extreme threat if all Soviet forces threatening NATO cannot be placed at risk or if NATO has any real expectation of being able to inhibit or challenge the extent of control of possible escalation the Soviets now have in the NATO European theater. The problem underlying the question is the continued wishful reluctance of the questioner to accept the facts of Soviet operational nuclear forces in the theater, the premise of the strategic concept (deterrence) or to agree on the selected means of achieving it.

Of far more relevance to the real problems are the corollary questions:

• Is it realistic to think that we can retain escalatory control of conflict in NATO or that reliance on external strategic nuclear forces will be effective to prevent major aggression in NATO Europe? *Answer*: Not unless NATO improves the quality and capability of its European theater nuclear forces and its external strategic forces. The loss of NATO's escalatory control in the event of European conflict (and such control is a valid tenet of the strategy, notwithstanding its difficulty) is the result of continued procrastination in responding to the full dimensions of the threat—in delaying the modernization and expansion of our conventional forces for direct defense; in failure to modernize and make more efficient, survivable, and capable the theater nuclear forces of the alliance, in an avoidance of developing the means of chemical warfare in the face of grave threat in this awesome area; in reluctance to address logistic deficiencies in almost every aspect of defense; and in the failure of the United States to expand, improve, and posture the external strategic forces at a rate relative to the relentless growth of the Soviet nuclear arsenal. It is not a matter of failure in the provisions of the basic strategy; it is a failure to provide the minimum forces to deny the Soviets control of escalation.

NATO's reasons for these failures are legion. Nevertheless, the result is now apparent: The Soviets have not only denied NATO conventional and nuclear escalation control in NATO Europe but also have established their own escalatory control near the borders of the Warsaw Pact nations. This fact is now being recognized by NATO and some belated steps have been initiated to redress the imbalance (for example, the long-range theater nuclear force modernization of the allied command for Europe with Pershing IIs and ground-launched cruise missiles) at least to the point of challenging the Soviet's escalatory control and beginning to place their second and third echelons of reinforcing forces at risk. These programs are late; remedial actions are costly in political capital, resources, and personnel; alliance support for improvements is spotty; the tyranny of minorities is plaguing decisionmakers in member nations and how much time is available for healing and catching up is uncertain. The decisions of the North Atlantic council on theater nuclear force modernization are far from implementation; the issue must remain on center stage until the actual operational deployments are completed. It will take extreme political negotiating skill to carry this critical force modernization program to fruition, but it must be done.

The loss of nuclear superiority and the degree of escalatory control it provided has reduced the readiness with which NATO would be prepared to initiate the use of nuclear weapons. In present circumstances, allied resort to theater nuclear weapons would be likely to leave the allies worse off if the Soviets responded in kind. The degree of escalatory control that came with nuclear superiority cannot again be attained; restoring general nuclear parity is essential if we are to deny escalatory control resting in Soviet hands.

Concurrently there must be recognition of the far greater significance of the conventional element of the deterrent than has ever before been the case. The Soviets would always have to weigh the risk of the West turning to nuclear weapons as a desperate last resort after a genuine, stubborn, and sustained resistance was overcome. But if NATO's conventional capacity to resist was markedly inadequate, for example a matter of days or at best a few weeks as is now the case, they might well come to a different conclusion. They could come to believe that a NATO that would not prepare to fight more seriously than that would be equally unready to accept the risks of nuclear war. In such a case the credibility of the deterrent would be dangerously low. Hence the need for strengthening the conventional forces of NATO is greater and more urgent than ever before.

The agony of attempting to devise, articulate, obtain support, and achieve a consensus on a new basic strategy for NATO is a candle that is not worth the game. The needed revisions are not ones concerning the concept itself. Rather, agonizing political decisions on priorities within and among NATO member nations are needed. And we must develop a consensus for expanding the application of the collective concept of deterrence and defense to vital interest areas of the allies beyond the political borders of member states. The immediate candidate is the external oil supply vital for Western Europe and important for the United States. Sources of other critical resources (metals, minerals and chemicals) must be protected to preclude economic chaos and regressions in the standards of living among the allies.

- Does not "forward defense" preclude militarily desirable concepts of "maneuver," "mobile reserves," etc.? *Answer*: Not in and of itself, but only if the United States defaults on its obligation to preserve political integrity for its allies. Although the recognized political imperative of forward defenses could in some situations inhibit optimum defensive positions (obviously it is easier to defend west of the Rhine than east of it), it is so understandable and reasonable that serious objection for purely tactical reasons is muted. It is not the basic strategy or the forward defenses that deny NATO maneuverability and mobility; it is NATO members' unwillingness to provide the resources—the forces, equipment, and logistics for the forces needed to enjoy the benefit of these time-honored and battle-tested tactics. The lack of mobile maneuver elements and in-theater reserve formations is not a fault to be attributed to strategy or to the field commanders. It is another instance of unwillingness to make proper and long-term political decisions to provide the essential equipment to implement the strategy we have devised.

- Is it reasonable to expect to fight with a truly collective NATO force, under international command, or will we field merely a collection of national forces? *Answer*: Probably the latter, although the strategy is built on the *collective force* principal except for logistic support, which is the responsibility of individual nations. NATO has developed the allied command framework for the cohesive operational control of forces provided by nations. This is a unique hallmark of the NATO alliance, setting it apart from the historic coalitions of the past and giving rare opportunity for a degree of credibility in the whole that is greater than its parts. This question goes to the heart of another serious failure on the part of nations to implement the NATO strategic concept—that is, a failure to man, support, and vest the integrated alliance staffs and commanders with the authority they need to be efficient. Fortunately this omission could and should be corrected rapidly with only minor investments of resources or manpower. Basically, it takes only a decision by nations to bring about the initial corrective action.

The allied command structure exists, but the allied commanders and staffs that will conduct the collective defense of NATO have little or no operational responsibility in peacetime and little or no voice in the disposition, composition, or logistic support of the forces they would command in war. This situation is and has been a gap in NATO's posture as a deterrent to war. It could prove to be a serious, possibly fatal defensive problem in combat due to confusion and disarray at critical junctures. The increased capabilities of Warsaw Pact aggravate the flaw for NATO and emphasize the need for prompt corrective action.

Member nations should insist on truly operational battle staffs (prepared to handle intelligence, planning, operations, and logistics) in the headquarters of NATO's allied international commanders and should readily make available the national staff officers now performing those functions in national staffs. We should undertake to support these allied commanders. The NATO nations should assign the operational control of more of their ready, in-place forces to the allied commanders; only air defense forces are presently under their operational control. It is an anachronism to think that national forces can fight effectively under diverse national commands in Western Europe, in the Atlantic, or in the Mediterranean. It is equally absurd to think that they can perform at optimal efficiency under allied command and staff direction if the first experience of such command occurs at the outbreak of war or the imminent threat of actual conflict. Those who think NATO now has that capability or who consider it of little consequence do not understand the situation.

NATO members should take the political decisions to correct this long-standing omission and start now in peacetime to build the kinds of allied commands and staffs we would want to employ our forces in war. This simple, low-cost improvement in NATO's deterrent capabilities is in keeping with the strategy of optimal deterrence and defense. Many other important things concerning NATO militarily will rapidly begin to fall into line as a result of this low-cost organizational improvement (for example, intelligence fusion; indications and warning; coherent tactical command and control communications; effective, timely dispositions; improved positioning and movement of resources; planned reception and support facilities; logistic understanding; tactical evaluations). This seemingly minor improvement is a departure from the NATO patterns of the past 30 years. Consequently it will not be an easy political decision but will have major military impact and long-term military implications, all good for our lagging deterrence and our inadequate defense as we move into an era of increased danger where Soviet military leaders might be tempted to push against a leaning wall.

- Is it realistic to assume that NATO's political authorities will be able to reach timely decisions for appropriate collective action in a crisis? *Answer*: This question is saved for last since NATO's capacity for crisis management is the most vexing problem in the alliance. The frank answer is "We don't know, for we have never really been tested." NATO has no intelligence except that which its individual member nations share with it; it has no intelligence-gathering assets of its own and, except for some spot tactical intelligence from the NATO air defense network, the alliance is dependent on the several member nations for indications and warning of attack. Aside from an actual attack and immediate response in the face of attack, the initial recognition of an incipient crisis for the alliance must be a national one; the indicators detected by nations and the military and political assessments made their leaders must be shared with the allies. The extent of sharing by the detecting nation (or several nations) must be so extensive that the allies must be convinced, to the point of political consensus, of the fact of crisis. Only then can NATO's preplanned, collective crisis management procedures come into play. Without political consensus between and among the capitals of

NATO members, there can be no full-scale alliance response; those nations convinced of the fact of crisis would be left with solely national measures of readiness or measures taken by a coalition of willing allies acting in concert but without the full-fledged support of NATO's crisis management procedures and measures.

It is expected that the all important initial warning of crisis will be a multifaceted process: nation to permanent representative, nation to major NATO commanders, capital to capitals, and capitals to permanent representatives. The process can be expected to be ad hoc, due to the unpredictable circumstances of crisis, however. In any event the decisions of nations must be communicated rapidly to their representatives to the North Atlantic Council in Evere (Brussels), for it is through council consensus that the NATO alert measures are implemented. NATO's preselected and precoordinated alert measures are extensive and critical to crisis management but not well understood at the national political levels where they must be authorized. Since such understanding is at the heart of the collective response principle of NATO's strategic concept, it is a capability that must be recotnized and mastered by the principals of nations in the North Atlantic council and national decision levels. If it is not better understood, constantly refined, and fully supported by individual member nations, both NATO's deterrence and its defenses will be weakened. This is another critical, no-cost capability immediately available to the allies to improve the credibility of NATO's deterrent posture; it takes study and understanding and deliberate revisions when the alert measures are found to be wanting or dated.

SACEUR General Rogers has petitioned for an improved, full-time indications and warning center in his intelligence staff to support nations and the North Atlantic Council in this critical aspect of our overall responsiveness to crisis—our credibility. General Rogers has given it top priority; it should be accorded equal priority for support by the United States and all our allies. While U.S. intelligence collection is not without gaps and uncertainties, it is the most comprehensive of any NATO nation, and NATO will not be well served without continued, unreserved sharing of U.S. indications and warning as well as U.S. support for all available avenues of making our best evaluations available to our allies.

THE IMPERATIVES OF A GLOBAL THREAT AND THE INHIBITIONS OF A REGIONAL NATO CONCEPT

Faced with continuous opposition to an extension of the NATO concept beyond NATO's geographic and political borders,[3] Ambassador Harlan Cleveland has repeatedly voiced the optimistic proposition that, when tested, our alliance was flexible enough for the formation of a "coalition of the willing" to respond to an external crisis affecting the alliance. That optimism may offer the best solution, for we do not have the luxury of approaching our diverse interests as a question of "either or." We must seek such coalition.

Although many of us share Ambassador Cleveland's optimism, such a coalition is at best an ad hoc arrangement, fraught with all the uncertainties of any last minute lash-up. Also, danger is ever present that, in the press for external support beyond the NATO area, we may be counterproductive and weaken the stability of that always critical area of confrontation.

As we enter an era when the vital interests of the NATO alliance so evidently extend beyond the potential borders of the member nations, the United States should be leading in an attempt to do as German General Gerd Schmuekle suggested in early 1980: "if we cannot take Allied actions in matters external to the alliance, we should at least act as Allies in such matters."

There is, of course, no assurance that our NATO allies will accept this linkage of their security interests outside NATO's classic boundaries. There can be no certainty that its allies will follow the lead of the United States in matters affecting the alliance or external to it, be they matters of concept or capability, no matter how legitimate the intentions of the United States. It is reassuring, however, to reflect on the many successful instances in which the U.S. position on matters affecting the alliance or even external to it was strong, consistent, altruistic — and successful! One could say with considerable historical precedent that once the allies have known where the United States stood, they have often been willing to stand beside us. The failures of the United States as leader in the major undertakings of the alliance seem to result from initiatives where our logic was unsound, our determination lacking, our articulation inconsistent, and our leadership weak and divided. The instances of success and failure should tell us something important: if we know what we are doing and have taken the allies into our confidence, and if we are *right*, we will find our reliable allies on stage with us and part of the coalition of the willing, not only in maintaining the deterrent credibility of NATO but also in obtaining the support so essential in vital areas outside the NATO area. But if we overplay our hand, ask too much, threaten too much, or if we are seen to be weak, internally divided, irresolute, and uncertain, we will not deserve the worldwide allegiance and support of our allies—and we probably won't get it!

It seems appropriate to conclude with these remarks made by Secretary of State Alexander M. Haig, Jr., who served as supreme allied commander, Europe, from 1974 to 1979:

> The current NATO strategy of flexible response is based on a balanced triad of conventional, theater nuclear, and [external] strategic nuclear forces, in which the deterrent value of each component is magnified by its relation to the other two. Through this linkage, the Alliance seeks to deny any aggressor confidence in his ability to forecast the form of the Alliance response.... He would be required to face uncertainties regarding how the Alliance would respond, where it would respond, and what level of conflict might ensure.
>
> *NATO's biggest problem, then, is not with its strategy.* Flexible response provides [the strategic framework for] a complete deterrent; it is a strategy with universal overtones. Politically, it represents the best compromise across the Atlantic that could have been devised. To tinker with it at this time would only create additional difficulties during a period of unsettled Atlantic relationships . . . Today, the Soviet Union's achievement of strategic parity or better with the U.S. has forced NATO to rely more heavily on the triad's other two legs—theater nuclear forces and conventional forces.
>
> The improvements in these elements of the triad cannot be token gestures. Theater nuclear weapons, long the ignored component, have assumed an increasing deterrence value . . . Similarly, the character and strength of the conventional posture of the Alliance are more critical than before . . .
>
> NATO strategy is sound but NATO is finding it difficult to provide the means to execute the strategy. *Rather than lose ourselves in the vagaries of a new strategic concept, we should improve the means to implement existing strategy* . . . But *the Alliance cannot dally in its assignment or priorities.* The security of the industrial democracies is crucial to national survival. Every dollar must be spent wisely and the NATO publics must be educated to appreciate the value of the expenditures. In this quest, *there exists no substitute for American leadership.* [All italics added.][4]

NOTES TO CHAPTER 7

1. Quoted by M.W. Hoag, *Foreign Affairs* (January 1958).
2. See North Atlantic Treaty, Article 6:

   For the purpose of Article 5 an armed attack on one or more of the Parties is deemed to include an armed attack

   - on the territory of any of the Parties in Europe or North America, on the Algerian Department of France, on the territory of Turkey or on the islands under the jurisdiction of any of the Parties in the North Atlantic area north of the Tropic of Cancer;
   - on the forces, vessels, or aircraft of any of the Parties, when in or over these territories or any other area in Europe in which occupation forces of any of the Parties were stationed on the date when the Treaty entered into force or the Mediterranean Sea or the North Atlantic area north of the Tropic of Cancer.

3. Though some latter day NATO practice would seem to accept constraint of the alliance's collective actions to a strict interpretation of Article 6 of the treaty, Ambassador Theodore C. Achilles, one of the original members of the treaty drafting group, has given important historical insight that is contrary to current proscriptions. His article in the October 1979, *NATO Review* is instructive; he writes:

   > The Tropic of Cancer was adopted as the southern boundary of the treaty area simply to avoid involving any part of Africa or any other of the American Republics (Mexico, Cuba or any others). It is worth recalling today that the treaty area, as defined in Article 6, is simply that in which an armed attack would constitute a casus belli; there was never the slightest thought in the mind of the drafters that it should prevent collective planning, manoeuvres or operations south of the Tropic of Cancer in the Atlantic Ocean, or in any other area important to the security of the Parties.
   >
   > The British, French, Netherlands and Belgians would have liked some commitment for assistance in the event of attack on their overseas possessions but realized that this would have aroused insuperable opposition in the U.S. Senate. Agreement was easily reached on Article 4 that "the Parties will consult together whenever, in the opinion of any of them, the territorial integrity, political independence or security of any of the Parties is threatened." It was understood by all that the scope of this Article was world-wide.

4. Quoted from an address at the Brussels conference "NATO, the Second Thirty Years," September 1–3, 1979.

8 RESOURCES AND REQUIREMENTS
Roy A. Werner

The relentless Soviet military buildup of the last two decades and the complexities of the international environment highlight the shortcomings in the present NATO defense capabilities. Yet the NATO countries' gross national products (GNP), military forces, and population are greater than those of the Warsaw Pact countries. The reality of its deteriorating position in balance conventional military power compels the West to improve combat capability and use resources more efficiently. Complicating the situation is the fact that the statesmen who must develop security policies and convince Western electorates of the merits of those policies lack the clear security imperatives of an earlier era. A gap exists between what is militarily desirable and what is politically and economically feasible.

The paradox of deterrence is that only by possessing military strength and the will to use it can a nation or group of nations mitigate against having to resort to force. But in all prosperous democratic societies, in the absence of clarion clear aggression, it is difficult to convince people there might plausibly be a threat to their independence.

THE SOVIET THREAT AND THE NATO RESOURCE BASE

The Soviet military buildup, nuclear and nonnuclear, since the early 1960s has surpassed anything previously seen in world history. The megatonnage, accuracy, and overall capability of Soviet central strategic nuclear forces are at least the equal of their U.S. counterparts. Soviet theater nuclear forces and land-based air forces, including the modern SS-20 missile and the high-capability Backfire bomber, are deployed in ways that threaten Western Europe itself, Japan, China, Korea, the Middle East and the sea lanes connecting these regions. Their general purpose forces have attained a first-class offensive capability. Furthermore, for the first time in history the Soviet navy has acquired and demonstrated a global naval capacity. In effect Moscow's emergence as a global military power necessitates a painful reversal of budget priorities in NATO countries: less money for social and economic programs, more money for defense.

Although Soviet military weapons may never be used, they affect perceptions and are an important currency in international politics. Budget allocations are of course political decisions reflecting societal preference patterns. There will certainly be, as there have been in Britain, critics who say we cannot afford to expand defense no matter how badly it is needed, because of the health of the economy. If the economic plan of the administration of President Ronald Reagan fails, in 1983 or 1984 when the military investment really starts to "take off," the Reagan administration will likely come under increasing pressure from constituencies and political allies to limit defense expenditures. Moreover, although the average American inflation rate was only 2 percent for the 16 years of the Eisenhower, Kennedy, and Johnson presidencies and only 5.3 percent for the first Nixon administration, the average for the second Nixon term plus Ford's and Carter's terms has been closer to 9 percent. In the 1950s and 1960s, productivity rose by 2.4 percent per year; during the first Nixon term, by 2.1 percent; in the mid-1970s onward by only 0.5 percent. In 1969 crude oil cost $3.21 a barrel. In 1981 it cost at least 10 times as much. Hence, the room for economic maneuver, for trade-offs between unemployment, inflation, avoidance of a social explosion, and defense is therefore much more narrowly limited than it has been for decades. This is true for the European allies as well as the United States.

The allocation of scarce resources requires deciding how much to budget for defense as opposed to other social purposes. Expanding the budget to include supporting allies, invites inevitable comparison of the efforts of those other states. Of course, aggregated micro-analyses indicate only the relative burdens since gross national product is not uniformly calculated in different countries and cannot adequately measure purchasing power in national economies or the benefits of technological or economic spin-offs. More importantly, military effectiveness, or the efficacy with which a nation generates military power, cannot be measured.

Burden-sharing, however, inevitably becomes a domestic political issue in each nation. Nevertheless, in the absence of mutually agreed upon divisions of labor and resources, there is no precise way of measuring appropriate contributions from different countries.

NATO summit conferences in London in 1977 and Washington in 1978 laid the framework for a policy of a 3 percent per year increase in defense spendings and the ratification of a long-term defense program designed to achieve greater interoperability and standardization enroute to a more effective coalition defense posture. If fulfilled, these programs will enhance considerably the combat capabilities of the NATO alliance. Whether they will be sufficient in the mid-term remains to be seen. There is strong evidence to suggest that a 3 percent standard increase may be insufficient to correct the military imbalance.

The "3 percent solution" reflected competing political realities. The desire to reduce tensions with the Warsaw Pact and spend more for domestic social programs competed with the equally obvious need to spend more money in order to strengthen and improve Western defenses. Committing governments to definite increases ran the risk of exacerbating volatile national debates over social programs versus defense if economies soured; this appears to have happened in Great Britain and may be occurring in West Germany. The future of the 3 percent commitment is in doubt given deteriorating economic conditions and the intense struggle in the democracies to decide government budget priorities. Whether NATO can continue to rely on the inadequate procedures of the past, given the Soviet military buildup is a crucial question. Indeed, the pressure for arms control negotiations in the West often come in part from a desire to put an economic cap on the defense effort in order to allow greater social programs. After all, peaceful Western electorates seldom rush to vote for the politician urging increased sacrifice on behalf of defense spending.

Debates about burden-sharing have been commonplace within NATO since 1960, when the chronic American balance of payments deficit first alarmed officials. These concerns led to a series of agreements between Washington and Bonn on German purchases designed to "offset" American expenditures to maintain U.S. forces in West Germany. In the early 1970s, after the dollar crisis and U.S. Congressional proposals to reduce U.S. troop strength in Europe, the debate again became wide-ranging. Unlike the earlier period, which emphasized balance-of-payments problems, the 1970s focus was on budgetary costs of U.S. forces stationed in Europe. However, attempting to assign specific costs, especially to indirect support functions, was difficult. Suitable criteria simply do not exist. Moreover, for the United States to maintain the same force structure domestically would probably be at least as costly but would degrade combat readiness and increase the need for additional strategic lift. Hence the only real option to reduce costs would be to demobilize troops.

As Table 8-1 shows, the American defense effort represents a higher proportion of its GNP than does that of NATO Europe. But the U.S. GNP is larger with a smaller total population. Further, economists would agree that nations with a higher per capita income logically may expect a relatively higher expenditure for defense. Thus some nations provide a larger share of collective defense than others. One possible alternative to such an arrangement in NATO would be to recast the military side of the alliance by creating a unified defense establishment controlling fiscal and force structure shares of a "security fund" to which nations might contribute an agreed proportion of their defense resources. Although such a proposal would maximize efficiency, it is politically unlikely.

What is needed and at the same time possible is recognition, on the part of our Western allies, that American global defense responsibilities indirectly benefit NATO. Thus, should it become necessary to withdraw personnel from U.S. forces in Europe to assist in an RDF deployment elsewhere, NATO must have ready plans to replace them with European reservists, firepower, and civilian assistance. Such a scheme also logically supplements NATO forces during war. Burden-sharing in the 1980s means devising plans that will enhance the collective security of all NATO members.

The NATO alliance apparently faces long-term competition with the Soviet Union. Western statesmen will continue to ponder whether sustained domestic support for increasing defense expenditures will exist. Nevertheless, the economic strength of the Western alliance far exceeds that of the USSR and its allies. Add to this equation a potential second front along the Sino-Soviet border, the promise of a retaliatory strike against the Soviet Union after a first nuclear strike, and superior combined Western naval assets. Considering all these factors the potential military might that could be deployed against the USSR over time is daunting. If NATO can defend itself against a sudden attack with minimum warning time, an ensuing war should ultimately allow NATO to prevail by converting superior economic resources into military muscle. The tasks for the 1980s thus are to continue the improvements already begun in readiness and to promulgate plans that will enable NATO to shift smoothly into a warfare mode if necessary. Neither of these objectives calls for gigantic surges in defense spending, but they do require increases, especially in weapons modernization, and, perhaps more important, coordinated and efficient planning. The issues become: How serious do the citizens of NATO states and their leaders believe the risks are? What are they prepared to do to cope with the situation?

Because governments desire stability, the tendency among the allies is to keep the proportion of GNP devoted to defense spending at roughly current levels and growth. However, equipment costs are rising faster than GNP or inflation. Personnel costs are generally rising as well. Even if real defense spending remains the same, the quantity or quality of equipment will shrink. Moreover, the scope for trade-offs between personnel and weaponry is more limited than many believe.

Automation of maintenance and operations increases capital costs. Extreme specialization can reduce duplication, but it is politically unlikely that any nation will forego its navy (for example) to enlarge its ground forces at the request of the alliance. Budget stringencies thus force NATO to seek efficiencies coupled with modest increases for procurement of weapons.

Table 8-1. Defense Expenditures and National Economies, 1979.[a]

Country	GNP U.S. $ Million	GDP per Capita U.S. $	Defense Expenditures Per Capita (U.S. $)	As a Percentage of GDP in Purchasers' Values
NATO				
Belgium	89,520	10,108	329	3.3
Canada	50,410	8,878	171	1.8
Denmark	50,410	11,521	257	2.3
France[b]	439,970	9,123	358	3.9
Germany, West	587,700	10,872	366[c]	3.3[c]
Greece	30,530	3,470	201	5.8
Iceland	1,880	9,906	...	...
Italy	218,320	4,834	116	2.4
Luxembourg	3,730	9,868	104	1.0
Netherlands	117,190	9,484	317	3.4
Norway	38,500	10,253	327	3.1
Portugal	19,540	1,877	65	3.5
Turkey	51,750	1,197	62	4.6
United Kingdom	281,090	5,626	267	4.8
United States	2,117,890	9,796	513	5.2
Warsaw Pact				
Bulgaria	28,450	3,200	81	2.1
Czechoslovakia	71,320	4,720	159	2.8
Germany, East	95,490	5,660	285	6.3

Hungary	36,860	3,450	84	2.1
Poland	128,330	3,660	99	2.4
Romania	38,170	1,750	57	1.4
Soviet Union	965,520	3,700	n.a.	11–14
Country	Population (000)	Active Military Forces (000)	Active Military as a Percentage of Total Labor Force	Reserve Military Forces (000)
NATO				
Belgium	9,840	107	2.8	54.4
Canada		79	1.1	19.1
Denmark	5,104	33	1.6	154.3
France[b]	53,278	578	3.1	350.0
Germany, West	61,310	492	2.5	755.0
Greece	9,360	180	6.5	290.0
Iceland	224	...	...	...
Italy	56,697	490	2.4	738.0
Luxembourg	356	1	0.9	...
Netherlands	13,986	107	2.7	171.0
Norway	4,059	40	2.6	245.0
Portugal	9,798	81	2.2	...
Turkey	43,210	698	4.3	425.0
United Kingdom	55,822	324	2.2	257.6
United States	218,548	2,027	2.9	818.7

(Table 8–1. continued overleaf)

Table 8-1. continued

Country	Population (000)	Active Military Forces (000)	Active Military as a Percentage of Men, Aged 18-45	Reserve Military Forces (000)
Warsaw Pact				
Bulgaria	8,814	150.0	8.4	240.0
Czechoslovakia	15,138	194.0	6.4	350.0
Germany, East	16,756	159.0	4.7	305.0
Hungary	10,685	104.0	4.8	143.0
Poland	35,010	317.5	4.2	605.0
Romania	21,855	180.5	4.1	502.0
Soviet Union	262,442	3,658.0	6.6	5,000.0

Sources: "Financial and Economic Data Relating to NATO Defense," December 9, 1980, NATO Press Service, Brussels; *The Military Balance 1979-1980 and 1980-1981*, International Institute for Strategic Studies, London; *The Europa Yearbook, 1981*, London. (The data for GNP are from 1978.)

a. The data presented for NATO are based only on the NATO definition of defense expenditures. Hence, national figures may differ because of definitions.

b. France is a member of the alliance without belonging to the integrated military structure; the relevant figures are indicative only.

c. These calculations do not include the expenditures on Berlin, which do not come within the NATO definition. If these expenses were included, the percentage would be $445 (per capita) and 4.1 percent.

THE CHANGED NUCLEAR BALANCE AND NATO'S FORCE POSTURE

NATO strategy formerly called for only short nonnuclear war against the Soviet Union in which American nuclear superiority need merely to have been alluded to in order to help end the war on satisfactory terms; failing that, it would have destroyed the aggressor. But in an era of Soviet nuclear theater superiority and strategic parity, this strategy is no longer a credible deterrent.

Inferiority in ability for conventional warfare is increasingly dangerous. Large, modern, and continuously modernized conventional forces, including reservists, are required. Both available conventional war reserve stocks of all kinds and a more reliable scheme to provide supplies for periods well beyond the current stocks, which would suffice for only a matter of weeks, are necessary if nuclear conflict is to be avoided. The concurrent modernization of NATO nuclear forces is necessary since the Soviets might not accept our doing well with conventional armaments in an actual war. They might resort to nuclear arms if they were losing; their military exercises routinely include nuclear warfare practices. Certainly we would not wish to accept defeat in a conventional war just because we had run out of iron bombs or gasoline for tanks and trucks.

Improvements in NATO capabilities seem to be underway. In 1980 an equipment program requiring substantial improvements in delivery systems and command, control, communication, and intelligence ($C^3 I$) systems was adopted. NATO's improving technical ability to acquire precise information about potential targets, guidance system improvements in optics and other sensing devices, the enormous increase in the ability to process data, and the development of precise means of delivering munitions should be further developed. How rapidly this technology will be exploited, however, depends on our joint will to commit the necessary resources.

The Warsaw Pact's quantitative advantage in conventional force is unlikely to be eliminated by NATO. Because of the classical advantages that adhere to defenders however, equality in numbers of men and weapons is not necessary. Traditionally military analysts have held that an attacking power requires a 3:1 attacker-to-defender force ratio to succeed. But if combat is analyzed in smaller sectors such as a corps area, the inherent offensive advantage of surprise could enable Warsaw Pact forces to achieve force ratios substantially above 3:1, thereby effecting a penetration and possible unraveling of NATO defensive lines. If NATO were unable to judge correctly where the main attack was being made and provide mobile reserves to reduce the force ratio, a grave situation would arise. The Soviets' doctrinal emphasis on surprise and shock, however, means they would most probably achieve local numerical superiority at their points of attack and then try to exploit these successes. Since the Soviets are generally equal in military technology and surpass NATO in *deployed* technology, NATO must increase its combat capabilities to maintain an acceptable balance. Despire Soviet gains the Western allies are still superior in general scientific and technical research. If this Western superiority could be exploited for military purposes it would remain unnecessary to match Warsaw Pact force levels to maintain an effective military balance. This is one of the crucial tasks for NATO in the 1980s.

In reality the military balance of the future will be determined less by the aggregate numbers of troops, tanks, and planes, and more by the successful application of the latest technology to the weapons systems of both sides. The advent of new technology to acquire targets, process information, and deliver munitions can help NATO to restore the conventional military balance and reduce reliance on the threat of tactical or strategic nuclear warfare that risks major destruction. Precision-guided munitions like the laser Hellfire, the Copperhead howitzer shell, and soon, terminally guided munitions such as the "fire and forget" Hellfire are survivable, jam-resistant weapons that can help to overcome numerical inferiority. New electronic instruments that enhance intelligence reporting and delivery accuracy are also important developments. The interrelation between tactics and technology in warfare is bound to affect the critical variables of space and time, firepower and maneuver, offense and defense. A problem remains in translating new technology into specific applications to the battlefield. To maximize the potential application of technology will require greater understanding between military and civilian leaders. Introducing anything approaching the optimal mix of new technology into the current force structure will require new investment. Advanced technology is expensive and uncertain, but there is little choice. For example, the armor disparity is one of NATO's main problems, and current NATO antitank weapons (excluding the tank itself) are vulnerable to suppressive fires, they are often immobile, and less effective against the new armor.

The United States thus needs to increase its research and development efforts, reexamine its doctrine and tactics alongside similar efforts by its NATO allies. They should think especially about reducing the costs of firepower by substituting other means of combat such as deception, surprise, and maneuver. The XM-1 tank is, without question, a superior weapon that can lead counterattacks successfully, but can enough of them be produced at acceptable costs? The XM-1 exacerbates logistical and transportation problems since it consumes 3 gallons of gasoline per mile, weighs nearly 70 tons, and can only be transported one at a time by the seventy giant C-5A airplanes. Smaller, faster, shoot-on-the-move tanks are also necessary to ensure sufficient tank assets. Likewise the TOW antitank missile, which requires continual tracking for more than 12 seconds, and the M113 armored personnel carrier should be quickly phased out and modernized systems more equal to the Soviet weapons and the military tasks at hand made available. An additional advantage of new technologies is that are more likely to destroy a military target rather than civilians and nonmilitary facilities.

Ultimately some form of European procurement agency will be needed. Working in conjunction with the United States such an organization would ideally be able to devise more efficient research and development programs that could eventually form the basis for greater standardization through co-production schemes. Gradual transformation into a pragmatic U.S.-European team for all phases of weapons procurement from research through production is probably the only way to gain increased military capability without exorbitant cost. Otherwise, efforts will resemble the saga of the 81-millimeter mortar. Europeans wanted to sell this weapon to the United States, and the United States wanted to buy it, but it has been held up by the fact that it exceeded health standards for noise. To date, the two-way street is only partially successful, as national differences and domestic politics handicap coalitions of interested alliance partners.

INDUSTRIAL MOBILIZATION POTENTIAL

The Defense Science Board reported recently that the American industrial base to increase production quickly in time of war is "extremely limited" and may be nonexistent without major new investments in production facilities and critical materials. Certain shortages today cut to the core of the industrial economy of the United States and will constrain the procurement of military hardware. According to one *Business Week* survey, key problems areas are: large forgings and castings, bearings, machining capacity, semiconductors and metals, as well as engineers, technicians, and skilled labor.

In fact, defense industries are hard pressed to meet even the regular ongoing requirements of the nation's military program. Well-defined requirements for a production surge, let alone capabilities, do not exist. Funds to implement increased production are not available. Moreover, we are nearly 100 percent dependent on foreign sources for critical materials like chromium and cobalt.

Some of the current shortages and probably all can be solved, given a few years of persistent effort. Because current military production "stretch-outs" have been in effect only a short time, some substantial increases could come quickly as industrial capacity has not fully withered away. For example, Navy F-14 aircraft are being produced at the rate of 12 per year, down from 60. The A-7 aircraft is down from 30 to 21, and production of the M-60 tank has been cut over the last 2 years from 129 to 45 units monthly. These rates could all be increased without too much difficulty.

Beyond the specific problems of a few production lines, however, the United States' military-industrial base is made up of more than 25,000 manufacturers and material suppliers, plus government facilities like munitions shops, shipyards, and materials stockpiles. Under the Defense Production Act of 1950 and the Defense Industrial Reserve Act of 1973, the U.S. Department of Defense (DOD) has the responsibility for insuring that this network is capable of meeting military production needs in any national emergency. The Federal Emergency Management Agency (FEMA) is responsible for determining societal allocations upon which DOD is dependent for resources and labor, but much of the planning and operational capability at FEMA, is embryonic at best. President Reagan has stated his intention of securing an additional $100 million to upgrade FEMA programs. Emergency readiness tests at the planning and control level reveal the need to fix exact agency responsibilities and develop plans beyond their present "conceptual stage." FEMA must be able to structure its plans and programs to coordinate economic allocation decisions in periods of conflict. The alternative is to place the authority for these decisions within the Department of Defense; in that event FEMA would become a civilian-oriented disaster relief agency.

This would not be a desirable solution although it would respond to defense needs.

The stockpiling program of the United States began in the late 1930s to establish an inventory of raw materials to be used in a time of war when foreign supplies were not available. A period of reevaluation commenced in the 1960s and 1970s, focusing increasingly on the concept of an economic, rather than a military, stockpile. In the last two decades, under administrations of both parties, there has been a progressive reduction in the quantities of materials stockpiled. Nevertheless America's growing import dependency and the possibility of future supply interruptions dictate a renewed emphasis upon stockpiling. In particular, supplies of strategic materials from potentially antagonistic or unstable countries should be secured. Even if the United States has the potential to become self-sufficient, as in aluminum production, the lead time and economic dislocations of doing so necessitate for stockpiling. More significantly, despite repeated recommendations in the 1970s the United States still lacks a comprehensive materials management policy to clarify these issues and provide for future needs. Without such a policy, the United States is vulnerable.

Under a gradual program of sustained buildup, a fully modern military industrial productive base could be created. The immediate need is not for production plants but for a coherent structure that reflects adequate planning, the fiscal incentives to support industrial involvement, and serious thinking about the nature of industrial preparedness. Although a modern industrial base could be created, it is doubtful that the base could remain abreast of technological change. Clearly, some governmental involvement in corporate capital investment planning (in exchange for financial considerations) is desirable in order to provide a modernized surge capacity over time, rather than a massive buildup of depreciating facilities that advancing technology will make obsolescent.

The old age of plants and equipment is perhaps the most severe bottleneck. Forcing new technology and manufacturing techniques into old plants and equipment is extremely difficult and economically inefficient. But much of the military production equipment that exists in the United States is owned by the government: At least 50 percent of the equipment of the top twenty military contractors is owned by the Department of Defense. Nevertheless the Pentagon does not have a modernization fund for its equipment, and there is virtually no incentive in defense contracts for companies to rebuild their own capacity.

America's aging and relatively inefficient industrial base is part of the reason for the decline of the United States in the world marketplace. It is seldom realized that U.S. defense capability and the credibility of the NATO deterrent is seriously weakened by this atrophy. Sixty percent of the equipment used to manufacture military hardware is more than 20 years old.[1] The U.S. Bureau of Labor Statistics has revealed that our productivity growth rates are currently the lowest of all free market industrial nations. Bluntly stated, our existing industrial base is not able to compensate for inadequate war reserve stocks or to produce sufficient material for wartime sustaining rates.

The existing defense industrial base is too small to meet immediate wartime demands and too slow to expand. American military forces cannot easily replenish probable combat materiel losses in the early weeks and months of a conflict. Even some 6 months after a declaration of emergency and mobilization, much critical equipment could not be produced in sufficient quantities to be helpful. The base dedicated to the purpose is too small because we have not devoted adequate resources to defense facilities for ammunition, heavy weapons, and rapid expansion capabilities. Nor have our European allies adequately stockpiled material or analyzed wartime conversion plans to meet the needs. Equally important, we have not analyzed well the potential of private industry to convert to wartime production requirements. We all realize that future warfare on any significant scale will require some degree of industrial mobilization. Yet nothing currently under consideration in the Congress or DOD will noticeably improve the situation.

In reality the entire concept of industrial mobilization has atrophied since World War II. To put things in historical perspective, mobilization for that war was a lengthy process of trial and error that ultimately produced 300,000 aircraft, 71,000 ships, and 86,000 tanks—the sinews of victory. Because of American success in achieving a fivefold increase of defense production within 2 years in that war we take it for granted that a similar industrial mobilization is possible today. The process of changing economic strength to military production is a complicated and uncertain chain of events the success of which is too easily assumed.

It is questionable whether we could achieve the same successful results today. Current sophisticated armaments require longer lead times to produce but may be destroyed more quickly and totally than comparable Second World War weapons systems. Unused plant capacity is less. We are more than 50 percent dependent upon foreign suppliers for nineteen of the forty critical raw materials essential to our $2.3 trillion economy (see Table 8-2. The only comparable shortfall for the USSR is bauxite and alumina. In a sustained crisis or prolonged hostilities, Western dependence on imports could be a serious handicap.

The need for industrial preparedness is clearly demonstrated by the historical record. Like today's short war scenario, the operating assumption prior to World War I was to use stocks during a short war and then rapidly demobilize. When events invalidated that assumption, U.S. forces were supplied by allies. Today, however, our allies are notably short of ammunition and other material. Indeed, some allies apparently believe that the United States will supply them in the event of a protracted conflict.

The World War I experience led to the National Defense Act of 1920, which charged an assistant secretary of war with preparing future plans. In the 1930s two industrial mobilization plans were developed—despite some opposition in the U.S. Senate—the 1939 plan recognizing that certain measures would have to be taken before the outbreak of war. Just as then, such planning is essential today to maintain American readiness. During World War II the Office of War Mobilization and the War Production Board allocated resources in accordance with defense priorities. This remains the only example of national industrial mobilization. The same concept of planned production was utilized during the Korean War, but a "cold base" (meaning that no production lines are operational) caused ammunition shortages and late material availability. During Vietnam War years, planned producers often had to compete on the basis of price with manufacturers of uncertain qualifications or experience or capacity to produce. Over time, therefore, the U.S. industrial base for manufacture of military equipment and supplies has been eroding. During the last 35 years there has been no real industrial mobilization. Both industry and the federal government have stopped planning for extensive mobilization of the civilian economy as conducted in the 1940s. In the Defense Department planning is limited to that required for the current authorized force (active plus reserve components) and to the portion of the privately owned base that, in conjunction with the government-owned base, is *actually* dedicated to or planned for the production of military goods.

Table 8-2. U.S. Net Import of Selected Minerals and Metals, Percentage Consumption, 1979.

Minerals and Metals	0	25	50	75	100	Major Sources
Columbium					■	Brazil, Thailand, Canada
Mica (sheet)					■	India, Brazil, Madagascar
Strontium					■	Mexico, Spain
Manganese					■	Brazil, Gabon, South Africa
Tantalum					■	Thailand, Canada, Malaysia
Cobalt					■	Zaire, Belgium, Luxemburg, Zambia, Finland
Platinum group metals					■	South Africa, USSR, United Kingdom
Bauxite and Alumina					■	Jamaica, Australia, Surinam, Guinea
Chromium					■	South Africa, USSR, Turkey, Zimbabwe
Asbestos					■	Canada, South Africa

RESOURCES AND REQUIREMENTS 227

Mineral		Sources
Tin	▭	Malaysia, Thailand, Bolivia, Indonesia
Fluorine	▭	Mexico, Spain, Italy, South Africa
Nickel	▭	Canada, Norway, New Caledonia, Dominican Republic
Potassium	▭	Canada, Israel
Gold	▭	Canada, Switzerland, USSR
Zinc	▭	Canada, Mexico, Honduras
Tungsten	▭	Canada, Bolivia, Korea
Cadmium	▭	Canada, Australia, Belgium, Luxemburg
Iron ore	▭	Canada, Venezuela, Brazil, Liberia

Source: U.S. Bureau of Mines.

The defense planning problem for industrial mobilization is exacerbated by arguments over long versus short wars. Originally this argument was over how to distribute limited resources. Providing for a short war was the basis. The goals were first, near-term readiness; next, mid-term modernization; and third, long-term sustainability. In the past these priorities made sense, but now, although the weight of evidence may support the likelihood of a short war, the prudent planner will hedge against such a possibility. More important, NATO may no longer be able to assume a short war without immediate resort to nuclear escalation, in which the Soviet response could damage NATO more than the original NATO attack upon Warsaw Pact forces. Therefore, shifting priorities to prepare for a longer term conventional engagement makes sense.

It should be recalled that before each world war many analysts said modern weapons were too powerful to allow for a protracted conflict. However tempting a short war theory is today, it could be just as incorrect again, assertions. Since NATO is now faced with nuclear parity and its own inferiority in some areas of conventional military capability, a more coherent defense effort is needed to restore the balance and allow the United States to cope peacefully with the dangers of the future. The bottom line is that we need modest but strategically significant increases in the resources devoted to industrial preparedness.

In an explosive international environment, military planning can no longer assume the luxury of enough time to gradually build up combat power. In late 1978 the U.S. Department of Defense conducted "Nifty Nugget" to test our ability to mobilize. That exercise confirmed a multitude of serious problems, especially in the industrial sector. As weapons and equipment became more complex, capital costs soared, equipment purchases were geared to just enough machinery to meet projected orders, and the fabled industrial base of World War II eroded. General Alton D. Slay of the Air Force Systems Command put the issue in perspective when he said, "I'm afraid that this time when we push down on the old gas pedal, that powerful eight cylinder response we've always received may not be there."[2]

The single most divisive issue in considering expansion of the industrial base is whether the next war will be long or short. In a protracted conflict, the only means to resupply combat materiel after stockpiles have been exhausted is through the industrial base. But in a short war the industrial base may be unnecessary or needed only to reach immediate stockpile objectives.

"Short war" advocates argue that industrial mobilization is unlikely and, in any event, would come too late to influence the outcome. The premise is that any war in Europe would be fought with the forces and stockpiles present; it would be an intense conflict, and resulting attrition and political realities would soon lead to a ceasefire. Proponents who envision a more protracted "long war," argue that we cannot safely assume a conflict with the Soviet Union will be of a short duration and therefore must prepare for a longer period of combat. In effect, the issue becomes whether the 1967 Israeli–Egyptian 7-day war or World War II are the more meaningful models of future conflict.

Such reasoning, however, is simplistic. First, the primary government responsibility is to defend this country. As a superpower the United States operates in an international environment opposite to a potential adversary whose defense effort has exceeded its by a considerable margin.[3] It is not reasonable to assume that at the superpower level a conflict will not be prolonged. Second, unlike the 1973 Middle East War, both superpowers are relatively independent of supplies from other nations in the early stages of war. Third, the possibility of military action in the Middle East with the U.S. rapid deployment force being engaged illustrates the heightened prospect of the United States being faced with a range of separate or simultaneous conflicts. It may be that enhanced preparedness for mobilization will provide an additional measure of deterrence. A more credible deterrent posture achieved through real potential to expand our industrial defense production can force potential aggressors to weigh the costs they could incur. Indeed, this policy option must be our course since democratic governments have never been able to compete with totalitarian regimes in the maintenance of existing military forces. Rather, we must rely upon our ability to mobilize and expand military forces in the event of a national emergency.

The case for greater industrial preparedness rests on twin propositions that every citizen can support. First, the U.S. primary global objective is to deter war. A revamped mobilization capability that makes clear to future aggressors that they will face the same consequences as First and Second World War aggressors should do much to disuade them. Second, even if deterrence fails, conventional forces supported by a viable industrial base and mobilization planning, can make it possible to refrain from having to escalate to the nuclear threshold. On the other hand, if U.S. and allied forces are considerably outnumbered, outgunned, and running extremely short of supplies and equipment, the policy options are bleak: either defeat or escalation to nuclear warfare.

PLANNING

Current planning assumes a requirement for X days of stocks on hand and an industrial base capable of achieving the required sustaining production rate within X days after the start of combat. It is doubtful, however, that sufficient stocks are available. Moreover, industrialists privately are skeptical that full wartime production rates in all areas could be achieved even within one year. A huge gap exists between requirements and production, and dollars, materials, and labor to close that gap are not available. Current plans do not define how the United States should achieve a state of increased preparedness, nor have rigorous analytical studies been done to clarify the options. The first task is to determine policy goals, define what is reasonably achievable under various incentive packages, and then develop and execute agreed upon objectives. Such a program will begin the process of restoring the industrial base as the arsenal of democracy.

Planning should provide for the development of mobilization production requirements for specific pieces of equipment and the identification of materiel and resource requirements for production targets. At present, however, because industrial participation is voluntary and unpaid, there is little basis for analysis that is trustworthy. Most significantly, the present planning system provides little insight into the potential strengths and weaknesses of industry, such as the pacing impacts of labor, parts, raw materials, subcontractors, or the potential for firms not currently engaged in defense production to enter the market. Production conversions need to be studied. How rapidly might a truck plant be converted to manufacturing tanks? What would be the human and material resources required? In effect, current industrial preparedness planning does not address the true potential of the industrial base to meet total mobilization requirements. A partial solution to the problem would be the appointment of a retired senior industrialist as a special assistant to the secretary of defense to assume responsibility for corrective actions.

Currently planning agreements are signed attesting to a firm's intent and manufacturing capability to meet specified schedules. There is no contractual obligation. Reimbursement for planning exists only when a reimbursable line item is in current production contracts. Because of this fiscal limitation, productivity enhancement or corrective measures proposed by contractors of government-owned facilities or private sector firms are rarely funded. Given this lack of funding and piecemeal planning, the actual state of the industrial base is further deteriorating over time.

Industry in fact has received disincentives to accompany the lack of positive incentives. Contracts during the Vietnam War were awarded on the basis of competitive bidding, thereby creating a disincentive and sometimes resulting in the relocation of government equipment to the new producer. It is also unclear, despite RAND studies,[4] whether an erosion of second-tier subcontractors has taken place. Low profit margins, foreign suppliers, lack of capital investment funds all lead to the withdrawal of producers of military equipment from the market. For example, more than 400 foundries shut down in the 1970s because of Environmental Protection Agency (EPA) and Occupational Safety and Health Administration (OSHA) regulations negatively affecting competitive market positions. Moreover, during periods of reduced defense spending and shifting defense requirements, not all planned producers can be assured of a specific level of business. In short, industrial interest in valid planning cannot be justified solely by the presumption of possible business contracts. Empirical evidence strongly suggests that a revised planning system and structured incentives are necessary before we can accurately assess the industrial base.

A new planning system must cope with trade-offs and production shortcuts that are achievable in a period of national emergency. The 1970 Joint Logistics Review Board Monograph on Logistics Support in the Vietnam Era reveals the dislocations created by low mobilization requirements as well as the problems involved in the reconstitution of the industrial base.

Current planning operates under existing OSHA and EPA rules, which would probably be modified during any military conflict. Maximum production now and during an emergency will therefore be different. The current peacetime oriented planning system fails to address the conversion of the economy as a whole to the production of war goods, the conversion of nondefense producers, and the state of manufacturing technology, especially in government-owned dedicated bases.

Present planning is imprecise, inadequate, and lacks any incentives to improve the process. Unlike the 1960s we cannot fiscally and politically afford an industrial planning objective of supporting 100 divisions. Instead the United States must create an industrial base that can meet peacetime requirements and provide a surge capability for "policy actions," "half-wars," or expanded defense readiness to take care of the needs of an all-out war. These situations could include

- Emergency foreign military sales to allied governments;
- More rapid U.S. modernization during a period of international tension;
- Replenishment of U.S. stockpiles during a crisis;
- Need to engage the rapid deployment force.

At the moment as one authority points out, "a low surge capability exists because we can only draw on material inventory already in the pipeline. Not only do we not have sufficient prestocked equipment in the quantities needed, we don't have the trained manpower to do the additional work, or much guarantee of expedited deliveries even using the Defense Priority System on a widespread basis."[5] The key to the problem is revised planning, the necessary prelude to any substantial investment and policy decision.

SUSTAINABILITY AND SURGE CAPACITIES

Although active military forces have basic weapons and equipment, adequate stockpiles of spare parts, major end items such as aircraft and tanks, and ammunition must be maintained if military forces are to be sustained in combat. A continuous topic of debate is the stockpiling of such materiel versus the investment required to maintain the desired state of readiness. The easily stated objective is to balance war reserve stocks against replenishment from the industrial base. Of course such analysis depends on the scenario and is conditioned by three variables:

- The duration of the war;
- The level of intensity of combat and the resulting impact upon consumption rates;
- Strategic warning time.

By varying these factors, specific requirements can be generated and priced. In this way decisionmakers judge the investment costs, cost effectiveness considerations, and the benefits versus the risks, and reach conclusions. The critical nexus is the rapid capability of the industrial base to gear up to wartime sustaining rates. This is especially important when one recognizes the persistent military planner's assertion that real combat requirements inevitably exceed planned requirements.

A major effort to measure the capability of the supply side of the defense market place to "surge" (defined as rapid and significant increase in production outputs) is necessary. "Nifty Nugget" exercise results in 1978 revealed serious problems in the adequacy of stockpiles and defense industries capability to surge. In effect, several months after war begins, U.S. forces would be constrained. However, the RAND studies referred to earlier suggest that with directed labor allocations and some additional machinery, planned producers could surge readily. Obviously, nondefense industries will have considerably longer lead times unless prior planning to minimize such conversions is undertaken.

Specific problems in developing a viable industrial base surge capacity include:

- Dependence on foreign suppliers of raw materials;
- Inadequate preparedness planning;
- Lack of economic incentives such as tax depreciation write-offs and investment tax credits;
- The costs of corporate borrowing and the high costs of new facilities;
- Declining productivity and lack of modernization;

- Absence of an allocation system to supply labor, material, supplies, equipment, and tooling in both declared emergencies and situations of lesser threat;
- The "on-again, off-again" nature of program commitments that are changed by the military departments of Congress; and finally
- Uncertainty over ability to utilize the U.S. allies' production facilities and stockpiles and the inability of the Atlantic allies and possibly Pacific allies to collectively structure their military procurement, research, and development programs to maximize their superior economic and scientific strength.

The relatively constrained investment in military production facilities since the onset of the Vietnam War has further exacerbated the situation. Modernization of the industrial base is essential if a real surge capacity is to exist during war. However, the Pentagon budget provides for relatively little modernization and industry has little incentive given the lack of tax benefits and the roller coaster nature of defense contracts. Consequently the understandable deemphasis of investment in the industrial base during the Vietnam years, constrained defense budgets in the 1970s, and the absence of incentives for commercial expansion have led to a shrinking of the defense industrial base.

A revised industrial base should identify, plan, and program for a "warm base" (production lines capable of rapid expansion) and modernized production lines for critical end items. Such lines could be kept operational through increased foreign military sales while also decreasing unit costs to all customers. National stockpiles of the necessary raw materials and long-lead-time components must be maintained. A new planning system to assess the situation adequately and to allocate resources to break bottlenecks must be designed. A variety of financial incentives need to be studied, including multiyear contracting to lessen program instability, larger profit margins, faster depreciation of capital investments to encourage new productivity enhancing investments, and a variety of tax incentives. Ultimately the system devised must be flexible enough to accommodate shifting defense requirements, changing defense spending patterns, and a wide range of contingencies.

Theoretically, there is a high correlation between capital investment and productivity. Data from the U.S. Bureau of Labor Statistics and the Organization for Economic Cooperation and Development show the United States lagging well behind other nations in these respects (see Table 8-3). This lag also partially explains the declining share of U.S. products in world markets. U.S. sales in high-technology areas such as aircraft (which dropped 8 percent in 1970-1980) and drugs (which dropped 12.6 percent from 1962 to 1980) dramatize this decline. Consumers notice it most sharply in the percentage of our domestic markets captured by foreign competitors (see Table 8-4).

Table 8-3. Investment and Productivity in the United States and Other Industrial Nations, 1960-1976.

Country	Percentage Increase in Manufacturing Productivity	Investment as Percentage of GNP
Japan	9.0	31.3
Germany	5.9	25.2
Netherlands	6.4	23.7
Belgium	6.5	20.5
Italy	5.9	21.0
France	5.7	22.2
Canada	3.8	21.5
United Kingdom	3.3	17.5
United States	2.2	14.2

Sources: U.S. Bureau of Labor Statistics and the Organization for Economic Cooperation and Development.

Table 8-4. Declining Percentage of World Markets Held by American Producers.

	1960	1970	1980
Automobiles	95.9	82.8	79.0
Calculating and adding machines	95.0	63.8	56.9
Electrical components	99.5	94.4	79.9
Industrial components	98.0	91.5	81.0
Metal cutting machine tools	96.7	89.4	73.6
Steel	95.8	85.7	86.0

Source: U.S. Bureau of Census.

The geographic remoteness and allies that previously gave America the time to prepare for war belong to a dead era. But, as Bernard Baruch said, "Modern war is a death grapple between peoples and economic systems, rather than a conflict between armies alone." Industrial preparedness may be the key to victory or defeat. Democracies cannot exist in "garrison states"; hence we must prepare now to expand military capabilities in the event of an attack.

MOBILIZING MILITARY PERSONNEL

During 1976-1980 the heavy losses of experienced technicians, officers, and noncommissioned officers, and shortages of reserve personnel placed U.S. military capabilities in serious question. Although pay raises and civilian unemployment have slowed the trend, the declining numbers of persons arriving at military age in the 1980s suggests further shortages of personnel.

The ability of the U.S. military to respond to a conventional war in Europe has been significantly reduced because of the decision of the U.S. government to adopt an all volunteer military force, as earlier chapters of this study reveal. Reliance has been placed on reserve units whose deployment schedules are nearly as demanding as those of the active military. Even so, there are serious deficiencies in reserve unit personnel and equipment needed to augment active forces upon mobilization. Many units are simply not combat ready because of these shortfalls.

The personnel policies of U.S. military service should logically support U.S. foreign policy and strategic objectives. But the discontinuities inherent in the all volunteer force structure would prevent a rapid and efficient flow of personnel into a combat theater in the early months of any conflict. In the event of a major conventional contingency, partially trained reserve units will merge with dual-based active units with equipment in Europe to reinforce our allies. Any outbreaks of fighting elsewhere in this world would only exacerbate the extreme personnel deficit. Moreover, the shortfalls in selected reserve units oriented for rapid deployment would increase pressure on pool of individual ready reservists (IRR). Yet in the army, for example, there would be a shortfall of 250,000 trained personnel even 90 days after mobilization. The recall of retirees from the preceding five years should only free a small number of personnel for replacement, and the selective service system is unlikely to be able to deliver personnel much before 4 to 6 months after the onset of war. (See Table 8-5.)

Fundamentally, there simply are not enough men and women entering reserve units of the IRR pools, and the machinery to deliver such trained personnel as are available would be insufficient for at least 120 days. Warsaw Pact forces enjoy the advantages of an earlier start on mobilization and because of their geographic location will have immediate reinforcement potential versus the U.S. requirements for overseas deployments.

The experience of three reserve callups in the United States since 1950 (Korea, Berlin in 1961, and Vietnam in 1968) reveals that unit readiness and personnel availability will fall well below expectations.[6] The result would be a prolonged delay in deployment or the cannibalization of other military units. Despite the admittedly important improvements in readiness and equipment made by reserve units and personnel in the 1970s, there are still serious deficiencies. According to the military's own readiness reporting system, many reserve units are C-3 or 4 (on a descending combat-readiness scale of 1 to 4) and thus are at best "marginally ready." In 1977 Congressional hearings the army reported that 43 percent of National Guard units and 54 percent of Army reserve units were rated "not ready."[7] Included in this assessment were a substantial number of early deploying units. Later the army reported to Congress that 36 percent of enlisted reservists were working in jobs for which they were not trained.[8] In addition, more than 49,800 reservists missed half or more of their drills in the first 6 months of 1979.[9] Hence, there

Table 8-5. Shortfalls in U.S. Military Personnel in the Event of Mobilization.

	Personnel on hand	*Personnel M + 90*	*Current Shortfall*
National Guard	378,400	440,700	65,800
Army Reserve	216,600	286,000	69,400
IRR pool	195,000	250,000-800,000[a]	
Inactive National Guard	71,000		

a. Varies by scenario; shortfall varies accordingly.

is historical and current evidence for doubting that the existing demanding deployment schedules could be met, unfortunately.

This analysis ignores reservists who would be medically unfit, hardship discharges, critical laborers, public safety and medical personnel who might be exempted, holdups because of equipment shortages, delays caused by transportation bottlenecks, and probable changes in deployment schedules reflecting battlefield needs. Almost as important is the fact that the entire "Total Force" concept of reservists and active duty reinforcement is critically dependent on two weak elements: strategic lift and efficient manpower mobilization.

Of course, similarly, personnel problems may limit Warsaw Pact effectiveness. Variables include an unknown number of troops that may be necessary for internal security in Eastern Europe; the reliability of Eastern European forces; the readiness, training, and equipment of Warsaw Pact cadre and reserve units; and the actual tactical plan of the attacking forces. Hypotheses and scenarios are debated endlessly, but the fact remains that NATO's initial ground combat capability is weakened by an ineffective system for mobilizing U.S. military personnel. Within the probable resource limitations of the NATO alliance, it would be prudent for civilian and military leaders to consider what steps might be taken to improve the initial personnel machinery for the conventional defense of NATO.

Possible alternatives are fairly clear cut: (1) maintaining the present relatively low degree of resource flows and commitment, thereby accepting the inherent risks involved; (2) bringing reserve forces up to the standards required; (3) compromising by developing an even more bifurcated reserve force structure of highly ready early deploying units and slower cadre units that would not be deployed until draftees began to join these units four months after war begins. Option 1 would continue the slow and steady progress of the 1970s but would buy relatively little enhanced capability. Option 2 would be the most expensive given the pay raises and bonuses that would probably be required. However, a return to selective service conscription or a special peacetime IRR draft would substantially reduce these costs if pay rates were below prevailing levels. Equally important, conscription would probably be perceived by both NATO allies and Warsaw Pact adversaries as a signal of resolve unequaled by anything short of the raising and deployment of additional active duty army divisions to Europe with increased aerial support and naval convoy capabilities. But the domestic political costs of such a shift are unknown. The question of fairness arises when a nation drafts only a small percentage of eligible youth. This issue is diminished but not entirely removed from the political scene by a draft lottery. Universal military training or national service schemes are popular with older Americans. On the other hand, these proposals could generate huge costs, unenthusiastic personnel, confront unions with the question of some job displacement of their members by cheaper labor, and philosophically retain the element of compulsion. A clear understanding of the NATO-Warsaw Pact force ratio imbalance and the Pact's geographical advantages make clear the necessity of an American reserve structure that can deliver trained personnel in the earliest stages of war.

THE AMERICAN STRATEGIC BUILDUP

The combined demands of improving the readiness, enhancing the reserves force structure, rebuilding the navy, modernizing the strategic and theater nuclear forces, and maintaining a rapid deployment force of reasonable weight, mobility, and flexibility—all these things will require difficult and courageous choices on the part of the public, the Congress, and the administration in Washington. They will require new military personnel policies and more dollars. They will require beefing up some production lines and steady, long-term defense planning and budgeting.

In the short term, given the objective of deterrence of the Soviet Union and Soviet-inspired military threats, the most efficient approach for the U.S. and its allies is not an overall, across-the-board general buildup. Rather, restoring the readiness of existing forces and building up the reserves are the key priorities since these actions will do the most to maximize deterrence and defense immediately.

In the next phase the best approach might be to concentrate on building up those things in which we excel and which worry the Soviets most. What is required is more effort and a conscious recognition of the value of facing the USSR with specific counters, such as limited-yield accurate missiles deployed in all theaters, supported by redundant communications, command, control, and intelligence (C^3I), and antisatellite weapons. It also means understanding specific Soviet attack designs and preparing to defend—and being seen to be prepared to defend—against them. For example, in Europe, since the

Soviets target our C^3I, we should beef up our C^3I defenses and provide the necessary redundancy (satellites, relays, microwave, and land lines) to insure continued operations. In all these areas both the United States and its allies possess strong productive capacity, personnel, and experience. High-altitude, large optics sensors based in space could provide near real-time surveillance, thereby enhancing strategic warning but requiring much improved antisatellite protection devices to avoid surprise.

THE INVESTMENT CHASM

The Soviet Union has been outspending the United States in the defense sector for some time. The cumulative advantage of that investment will be a flood of new equipment that the United States and its NATO allies simply cannot equal even if they desired to, for several years. Monetary investments in defense alone will not provide the answer to NATO's defense needs, of course. Although NATO economic muscle exceeds that of Warsaw Pact, the allies' investments are not fully beneficial because of conflicting national priorities and some duplication.

How to narrow this difference is one of the principal defense resource issues facing NATO. Indeed, if the Soviet system and economy are failing, the danger may be even greater that Moscow will seek to counterbalance these deficiencies by emphasizing sheer military strength. The single most important task is to get both America and its allies to do more and to do it more effectively. The economic burdens of defense mean that Western security cannot depend on unilateral American initiatives to respond. Rather, the strategy must be to create multilateral responses that maximize the alliance's total resource contributions.

Collaboration in the production and procurement of armaments is improving, but American modernization of weapons is handicapped by inadequate NATO infrastructure fundings. Host nation support (HNS) agreements that seek to identify and plan for civilian augmentation short of the front lines in the event of war need much more detailed planning and identification of specific personnel and materiel. Beyond the realm of weapons, logistics, training facilities, schools, and rethinking of strategic policies all affect additional avenues to curtail waste and enhance the allied defense product in an era of constrained resources. Likewise more must be done to confront the Warsaw Pact nations and especially the USSR with a credible dual front problem. Perhaps the single most effective quick addition to allied capability would be the emergence of some kind of NATO–Japan–China security relationship.

Finally, political leaders must recognize that the common good may require a pruning of the regulations, laws, policies, and practices by which each NATO nation seeks to maximize domestic production and employment while ignoring the most cost-effective means of satisfying the over-all alliance defense needs. Indeed, this problem may be most acute in the United States, given the constitutional separation of powers and the Congressional focus on geographic factors in the production process. But alliancewide this erosion of potential resource maximization is a serious problem.

NATO Standardization, Rationalization, and Interoperability (SRI)

The conventional force weapons and equipment disparity of NATO when matched against the Warsaw Pact illustrates a fundamental alliance resource problem. How does a democratic alliance equal the efficiency of a totalitarian alliance? The answer is clear: It does not. What then matters is the defense outputs from the production process. Granting national differences, NATO allies must still seek greater efficiencies. The process of seeking these economic gains is commonly but ambiguously referred to as standardization, rationalization, and interoperability (SRI). An agency to achieve these goals has long existed within NATO. The earliest attempts at standardization concerned 7.62-millimeter infantry ammunition in the 1950s; hence the quest for SRI may be compared to the multiple level Japanese game of *Go*, a bewildering series of interconnected linkages.

What is less commonly recognized is that these decisions are *strategic* production decisions. Once made, they will affect the military capability and perceptions of military balance. Research and development is a long-term military competition conducted amid great uncertainty. Obviously NATO does not know the impact of research and development decisions on Soviet perceptions, but NATO's own experience suggests that it may be as important as the number of weapons arrayed against each side. The unit costs of major military hardware items has been rising faster than real GNP or defense budgets. A self-correcting mechanism must begin, however, if NATO is to avoid putting all available dollars and efforts into a few very expensive weapons systems such as today's XM-1 tank, the inside of which resembles the sophisticated aircraft of a few decades ago, and a few very expensive support facilities. Simply put, numerical sufficiency must be as important as qualitative superiority.

NATO needs the potential economic benefits of SRI to narrow the Warsaw Pact advantages. Although nations share an interdependence in security matters, defense procurement inevitably involves domestic political and social goals that may not be compatible. NATO economic and military efficiency will never equal the Soviet-imposed standardization of the Warsaw Pact. Moreover, nations do not procure armaments solely on the grounds of military efficiency as any student of Congressional budget decisions can attest: Where the item is manufactured, the extent of employment to be generated, how many subcomponents might be scattered in various districts to effect a strong coalition for the procurement decision all weigh in the decision process. Rationalization to achieve more effective forces from the same funds will seldom be achieved without extensive political and fiscal sacrifices by individual states. Even then the effect will be less than maximum efficiency for NATO. This is not inherently bad, however, for NATO rests ultimately upon voluntary political cohesion, which in turn is dependent upon the economic well-being and social stability of each member state. Some efficiency loss therefore is to be tolerated.

The essential question for NATO thus is how much loss is acceptable to maintain cohesion while also improving allies defense capabilities? Differences in language and tactical doctrines further exacerbate the problems of alliance cooperation. It seems that progress must be made on a piecemeal, case-by-case basis, with the objective being improved efficiency. The American, British and German armies are already discussing integrated tactics and doctrines, common weapon systems, ammunition interoperability, and joint test standards for development. NATO-wide quadrapartite agreements already exist. A joint German-American logistics manual has been written. Also, the F-16, NATO AWACS, and AIM-9L air-to-air missile are now in co-production in Europe, and the Belgian MAG-58 tank machine gun, German 120-millimeter tank gun and Belgian squad automatic weapon will soon be produced in the United States.

Still, these are only first steps. Interoperability on the battlefield is the key to success, and that has yet to be achieved. Memoranda of understanding exist approving a common artillery round, except for tanks and mortars, and NATO has agreed to standard small arms ammunition. Some older production models are in co-production: the M113 personnel carrier, M109 Howitzer, improved Hawk missile, and radar systems. However, the Roland air defense missile system keeps getting reduced by American budget cuts and the multiple launch rocket system, agreed upon by Germany, Britain, France, and the United States is still in the acquisition process. Moreover, SRI considerations appear to be lengthening deployment times. Although the demands of threat analysis dictate faster procurement, the two-way street of SRI adds time to the procurement cycle. New NATO procedures for looking at weapons requirements and the family of weapons concept could solve this problem by apportioning development tasks earlier in the life of a new weapons program.

Ideally, over time, optimizing procurement costs through the utilization of only one or two production lines may come about. Then criteria emphasizing cost, performance standards, offsets, technological edge, and efficient production will determine suppliers. Thus far, little actual avoidance of duplication has occurred. A 1978 Congressional report estimates that approximately $3 billion annually is available as potential savings. This is less than two percent of the current alliance budgets.[10] Partial successes can be achieved through political will and economic cooperation. This remains NATO's procurement challenge.

THE FUTURE

Given the existing military imbalance between the East and West, the most important area for resource commitment for the West lies in improving the conventional balance. This does not necessarily mean massive increases in force levels or weapons. It does mean using existing resources intelligently and more effectively, with an infusion of modernized weapons. Already corrections have been made to NATO maldeployments that existed for a quarter century, some degree of interoperability is emerging, and there are examinations of the duplication entailed in each nation's maintaining "balanced military forces," and its own procurement and production organizations.

Nevertheless many still ponder whether a conventional defense by NATO is feasible at an acceptable cost given overwhelming Soviet superiority in almost every weapons system. In reality NATO has no choice if we are to survive in an era of strategic nuclear parity. If a conventional defense is feasible, it raises the nuclear threshold, improves deterrence of conventional attacks, and probably lessens political aggression. To fail to improve conventional forces is to accept a choice: surrender or suicide.

Obviously NATO allies have neither the political will nor the economic capital to achieve quantitative equality with the Warsaw Pact forces. The premise that NATO cannot defend itself conventionally rests primarily upon the size of the Warsaw Pact conventional forces. These "bean counts" of divisions, weapons, and aircraft are but partial indicators of combat effectiveness. Numerical assessments do not include important nonquantifiable factors such as strategy, leadership, intelligence work, morale, and quality. For that reason such counts cannot predict with accuracy the outcome of a battle or war. Rigorous comparisons of military capability are impossible because of force structure asymmetries, differing economic systems, and an inability to compute realistically or compare the rates of exchange of firepower between weapons systems. Although NATO needs to improve force readiness, its forces are not grossly inferior to those of the Warsaw Pact.

The era of strategic parity demands that Europeans contribute more to the alliance in conventional defense. Past metaphors such as *tripwire*, *firebreak*, and *ladder of escalation* obscure rather than define the problem. The fact is that Europeans spend a much smaller fraction of their GNP on defense than the United States does, although there are European contributions such as conscriptions that are not included in the dollar costs. During the 1970s the European contributions to the worldwide expenditures of NATO climbed from 33 percent to 45 percent. Moreover, net alliance costs make it cheaper for Europeans to assume greater conventional roles compared to further United States efforts. Periodic pleas have obscured the original mutual defense notion. As a result the relative shares of the defense burden today are still unfair. In addition, it is unlikely that American economic problems and the necessity of creating a viable military presence in the Middle East will allow U.S. initiatives to redress the shifting balance.

What is more significant is the political utility to be derived from Soviet superiority at the continential level and global perceptions of Soviet military prowess. The key issue remains whether the Soviet buildup has tilted the balance of power and endangers NATO security. A more decisive superiority would be required before the cautious Soviet leadership would threaten the status quo. Even so, the shifting balance of power does cause serious concern over future trends. Assumptions that the threat is diminishing are invalid. Prudence dictates that the alliance weigh this adverse shift and devise a common strategy to continue deterrence and defense. This may be especially necessary when one considers that a time of leadership transition is near in the USSR. Senior Soviet leaders in the 1990s may have fewer concerns over the human costs of war as World War II personal memories fade. It is, of course, inescapably cheaper in both blood and treasure to deter a potential aggressor than to fight.

Still, threat perceptions of the USSR have dimmed with the passing of time for many Europeans and Americans. Even the brutal invasions of Afghanistan and Czechoslovakia are but internal bloc affairs and protection of a client buffer state to many observers. Furthermore, Soviet willingness to negotiate meets the yearning of many in the West for a stable peace.

Undoubtedly few people believe the Soviets are soon going to invade Western Europe. But the absence of a direct and immediate threat of aggression in light of the disproportionate Soviet military buildup and modernization is no cause for complacency. There is, after all, no historical record of Soviet behavior under conditions of Soviet global military superiority. Moreover, NATO's inability to act in a coordinated fashion outside the North Atlantic treaty region presents the USSR with targets of opportunity elsewhere. Unilateral American counteraction in these situations will only degrade NATO readiness, the European sense of security, and may serve over time to drive a wedge in the alliance as American taxpayers perceive themselves to be carrying an unjust share of the total defense burden. For despite the shortsightedness implied, if the United States becomes unilaterally engaged militarily in the Middle East, a unilateral American foreign policy will be the aftermath.

The Hobbesian jungle of international politics demands that NATO increase its defense burden to move toward a more acceptable conventional balance with the USSR. Economically, alliance members need to develop mechanisms to exploit their greater strength and technical advantages. This is particularly true with respect to energy sources, weapons development, and stockpiles of raw materials. The future demands more, not less, alliance cohesion.

NATO, in short, needs a credible strategy backed by a commitment of resources to effect that strategy. The economic resources permit a greater defense effort, but only politics can compel such an effort—or at least make it possible. These are not problems of production or organization but problems of choice. The intangible factor of political will is the essential ingredient of military strength. To the extent that NATO can mobilize that willpower, it can and will maintain the resolve to defend itself.

The root of the public debate lies in the term *afford*. What precisely does it mean to "afford" increased defense efforts? Three meanings suggest themselves, but they are seldom delineated. First, has NATO the resources to produce the necessary armaments? Second, are increased defense efforts more important than other social and economic claims on resources? Third, are NATO nations prepared to forego desirable social programs to strengthen defense measures? Obviously the answer to the first question is affirmative. The second question can only be answered by individual leaders after weighing competing social and economic claims. Moreover, it must be remembered that budget reductions in the social field do not automatically lead to increased defense budgets. Only after the second question has been answered by national leaders can the NATO alliance begin collectively to consider the third proposition. Economic growth and increased defense efforts are not incompatible. Much depends upon international developments and how defense efforts are to be financed. Phrased differently, the choice is between devoting this decade to achieving higher standards of living and preventing war through cooperative efforts, or of enjoying slightly higher economic growth and then possibly losing all in the turmoil of war. As the history of NATO reveals, such problems exist not to be solved, but survived.

NOTES TO CHAPTER 8

1. *New York Times* (September 27, 1980): 1. The equivalent German and Japanese figures are 12 and 10 years respectively.
2. *New York Times* (September 27, 1980): 1.
3. "During the 1970s, Soviet spending on things related to military research and development, military weapon systems acquisition, and military facilities, exceeded that which the United States spent by $240 billion dollars. Last year the Soviets spent $50 billion more on these items than we did...." General Alton D. Slay, before the Industrial Preparedness Panel of the House Armed Services Committee, 96th Congress, 2nd Session, November 13, 1980, p. II-2.
4. See Geneese G. Baumbusch and Alvin J. Harman, *Peacetime Adequacy of the Lower Tiers of the Defense Industrial Base*, R-2184/1-AF, November 1977; Geneese G. Baumbusch and Alvin J. Harman with David Dreyfuss and Arturo Ganadra, *Appendixes to the Report on the Peacetime Adequacy of the Lower Tiers of the Defense Industrial Base: Case Studies of Major Systems*, R-2184/2-AF, November 1977; Geneese G. Baumbusch, Patricia D. Fleischauer, Alvin J. Harman, and Michael D. Miller, *Defense Industrial Planning for a Surge in Military Demand*, R-2360-AF, September 1978; and M.D. Miller, *Measuring Industrial Adequacy for a Surge in Military Demand: An Input-Output Approach*, R-2281-AF, September 1978.
5. General Alton D. Slay, before the Industrial Preparedness Panel of the House Armed Services Committee, 96th Congress, 2nd Session, 13 November 1980, p. III-18.
6. For a fuller explanation see, Roy A. Werner, "The Other Military: Are U.S. Reserves Viable?" *Military Review* 57 (no. 4, April 1977): 20-36.
7. Military Posture and HR 11500: U.S. Department of Defense Dod Authorization for Appropriations for Fiscal Year 1977, Hearings before the House Armed Services Committee, February-March, 1976, pp. 560-561. See also U.S. General Accounting Office, Letter Report LCD 77-442, December 21, 1977.
8. Ibid.
9. General Accounting Office, *Army Guard and Reserve Pay and Personnel Systems are Unreliable and Susceptible to Waste and Abuse*, 28 January 1980, pp. 24 and 31.
10. *NATO Standardization, Interoperability and Readiness*, Committee on Armed Services, House of Representatives, 95th Congress, 2nd Session.

9 PRIORITIES FOR THE FUTURE

George M. Seignious II

This book has reviewed the many problems facing the North Atlantic alliance. The Soviet threat has been identified as political as well as military, as reaching beyond NATO as well as threatening the homelands of the allies. The military needs of readiness, reinforcement, and resupply and of the power balance between East and West have been defined. The problem of resources has been analyzed. The nature of issues that can split the alliance from within have been outlined; and the urgency of rebuilding for the long haul, the conviction of the basic complementary values of liberty and responsibility to one's fellow man has been made apparent. It remains now to outline briefly the objectives that could result in an improved and strengthened NATO deterrent if given priority.

The paramount objective of the NATO alliance is to keep the peace while preserving freedom. In this context military potential is employed to attain the peaceful political end of deterrence. The essence of this defensive alliance is at heart political. It depends on the element of political will of the nations involved to work together in order to achieve the collective strength and coordinated policies necessary to deter aggression.

In 1947, when he launched the concept of the Atlantic alliance, British Foreign Secretary Ernest Bevin called for "such a mobilization of moral and material force as will inspire confidence and energy within and respect elsewhere." A blend of political and military measures, then, is essential for a healthy alliance. Priorities for this alliance cannot be categorized on any absolute or sequential basis. They must reflect, rather, a synthesis of various steps that are mutually complementary and that must be attained more or less concurrently.

POLITICAL OBJECTIVES AND PROBLEMS

The political priority of NATO in the 1980s is alliance unity, for without it every other program that we undertake to improve the alliance would be undermined or, in some cases, impossible to undertake in the first place. Indeed the greatest and most likely challenge to NATO in the decade to come may be not the actual outbreak of war and military action in the region but, rather, the threat of disunity fostered by domestic events within NATO nations, by Soviet activities, or by crises in areas outside of Europe, particularly in the Middle East but also in other parts of the developing world.

Of paramount importance in this regard is the strength of the transatlantic bond. The Soviets seek to divide the NATO alliance, and they can plow a fertile field on issues such as arms control versus force modernization, détente versus linkage, Mideast policies, energy and resource policy, and trade. The continuing Soviet effort to divide the alliance comes at a time when the combination of greatly increased Soviet strength and diminished confidence in American leadership since the 1970s presents new opportunities for troublemaking and divisive moves by the USSR. The allies should be prepared for major pressures on the first three issues just noted.

Force Modernization versus Arms Control

The potential clash between the programs of force modernization and arms control is highlighted by the issue of whether to deploy the long-range tactical nuclear force (LRTNF) weapons in Western Europe as they become available or to defer that deployment if there is not yet agreement on arms control measures. The Soviets, together with an increasingly vocal segment of antinuclear opinion in Europe, of course, press for the latter course of action, which would perpetuate Western theater nuclear inferiority. Thus during the 1980s a major factor making the search for cohesiveness between the United States and its European allies especially difficult may be a growing difference in emphasis placed on defense and arms control. During the 1970s the United States and Europeans both put great hopes and effort in support of arms control. It is fair to say that for the Carter administration it was in the forefront of national security policy. This preeminence pleased our Scandinavian allies and gained strong support from the Benelux countries and West Germany. The United Kingdom loyally supported this emphasis. It is possible that the disappointment if not disillusionment created by Soviet behavior and the lack of substantial reductions achieved by arms control could lead the Reagan administration to give such preeminence to force modernization that arms control measures might become subordinate objectives on the national security agenda. The Reagan administration has avoided the danger of sacrificing either of the twin tracks of modernization and negotiation. Support for both tracks will continue to be essential if the present level of support on both sides of the Atlantic is to be maintained.

Détente and Linkage

Similarly, the American emphasis on linkage between Soviet conduct throughout the world and East-West relations at times contrasts with the European tendency to attach great importance to the maintenance of détente between the Western European allies and the Warsaw Pact members. The Harmel report, adopted by NATO in 1967, made clear that defense and détente both have a role to play in achieving comprehensive security. But the Harmel report does not speak of détente as an end in itself, though that unfortunately became the popular impression. Rather, it refers to the search for progress toward a more stable relationship in which the underlying problems can be solved.

The European allies continue to attach great importance to the improved contacts with the Eastern bloc, the economic benefits of trade across the borders, freer movement across East-West boundaries and the stabilization of the borders, and are loath to risk sacrificing these gains over what they consider non-NATO issues with the Soviets. While the Europeans increasingly are recognizing that they can no more prudently avoid taking Soviet conduct outside the

NATO area into account, and the United States is increasingly conscious that it should not reject détente as a process just because it is not a helpful end in and of itself, this is an area where identity of point of view among the allies is not always readily obtainable.

Arms Limitation

More difficult is the question of arms limitations talks. The forthcoming antiballistic missile (ABM) treaty review places pressure on the Reagan administration to get its offensive strategic arms limitation house in order prior to reviewing and quite possibly altering the 1972 agreement on defensive strategic force limitations. Furthermore, the December 1979 dual track NATO decision on long-range theater nuclear weapons contemplated that the arms control component take place in the context of a strategic arms limitation treaty (SALT). In the absence of SALT II talks, the Europeans could become restive regarding the lack of progress in establishing a meaningful negotiation process on "gray areas" or long-range theater nuclear systems if some satisfactory substitute formula is not found. The complexities of structuring a negotiation on long-range tactical nuclear forces are tenfold more severe than seeking limitation on central systems for a number of reasons. The weapons systems themselves are more ambiguous (conventional/nuclear, role, range, verification) than the strategic forces included in SALT, and the concerns of the European NATO countries must be taken into account to a greater extent than with SALT. To overcome these complexities before achieving some formalized agreement on a SALT treaty will be difficult but imperative. Expectations by our European friends must be tempered by reality so that failure does not carry with it the seeds of self-destruction.

In addition, other reasonable arms control measures, both conventional and nuclear, holding open opportunities for negotiation, deserve to be examined. For example, fresh emphasis on confidence-building measures in the mutual balanced force reduction (MBFR) in Vienna or, as now seems more likely, under the French proposal for a conference on disarmament in Europe (CDE) may have a greater chance for success than a continuance of the reduction of forces track. The clear imperative for a dialogue between East and West recommends a persistent search for ways to make arms control a continuing and balanced part of our national security policy and the integrated policy of the alliance.

PROBLEMS OUTSIDE THE NATO AREA

What is now well recognized as a potentially sensitive area for NATO in the 1980s has to do with alliance response to pressures outside the allies' own countries. Because of the strain these pressures place on U.S. defense resources and because some of the outside problems, such as oil supply, pose fundamental threats to the security and economies of various NATO nations, these pressures promise to affect NATO unity more than ever before. Strong pressure on the European allies to carry more of the defense burden as the United States readies a capability in the form of a rapid deployment force and other priority programs will inevitably stress U.S.-European relations. And the need to work out common positions on issues that vitally affect NATO but that are outside its charter will strain the political resources of alliance members.

The twin crises in the Middle East—Afghanistan and Iran/Iraq—have elicited a loosely coordinated NATO response at best. The question remains whether NATO will engage in a rethinking of alliance strategy to cope with threats to NATO members that originate outside the formal boundaries of NATO. Alliance "crises" over extra-alliance matters have a long history, going back to the Suez Canal in 1956. The question facing NATO in the 1980s is the extent to which common policy may be developed to deal with global events of concern to NATO members and what the implications of such a strategy might be for division of labor within NATO. Such a strategy need not include formal NATO military commitments to regions outside NATO but could address such issues as distribution of resources within the alliance, and political-economic as well as military measures that the respective allies could take individually or in concert with other willing members.

Toward a Common Approach

In the face of these developments, the NATO allies cannot continue to be content with random, uncoordinated ad hoc responses, which themselves can constitute challenges to the unity of the alliance. There must be a common effort to maintain and strengthen alliance unity. In this sense, *unity* should not be taken to mean identity of views; rather, it should connote a continued purpose to subordinate such unavoidable differences as could affect the collective security of the Atlantic nations. There should be a politically coherent program that anticipates divergence and lays out an overall plan for dealing with troublesome issues. The intent of such a program would be to avert potential crises within the alliance or at the very least to mitigate the deleterious consequences should serious differences nevertheless emerge, as must be realistically expected.

In more positive terms, the thrust of this political program would be to identify each member nation's vital concerns and interests and, through consultation, find the overriding areas of common interest. As a first step in this process, a concentrated effort should be made to bring into closer focus the common threats to the alliance. In a political, economic, and military context, the threat must be evaluated comprehensively so that the security interests of the several states become realistically identifiable. In the past, the absence of an agreed assessment of the threat has helped to inhibit decisions needed to keep the alliance strong. If current threat assessments can be formulated that gain an increased element of commonality, then the risk of political estrangement within the alliance would be lessened.

It must be accepted as unlikely, however, that complete agreement within the alliance can be reached on all issues. It is a reality that the positions of each country vis-à-vis the East are not the same, and consequently the threat may not be viewed as universally unambiguous. The accommodation of differing perspectives constitutes one of the most demanding political challenges facing the alliance in the years ahead. The achievement of this goal will require a delicate balance between leadership and political sensitivity on the part of the United States. It will require on the part of all allies guarding against unilateral assertions and actions that unnecessarily aggravate or create conflicts, substituting an approach that recognizes and gives adequate weight to the views of our friends as well as our own. This will not be easy, but it would serve the common goal: survival as free nations.

DEFENSE OBJECTIVES AND PROBLEMS

It is not unnatural that an alliance devoted to collective security should find that many defense measures, such as doctrine, logistics, procurement, force structure, and burden-sharing are each in its own way highly sensitive to political considerations.

The Loss of Nuclear Superiority

The first of the military issues that must be addressed is the impact on force requirements of alliance doctrine and the strategy of forward defense in light of recent developments. Alliance doctrine has historically been circumscribed by the geopolitical reality that the Federal Republic of Germany requires a forward defense strategy. In addition, as demonstrated in the preceding chapter, the military reality that strategic nuclear parity does not any longer afford a realistic, distinctive, and separate deterrence has not been fully assimilated in the alliance. As the Soviet Union continues the buildup of its theater nuclear capability without offsetting moves by NATO yet accomplished and in place, the range of doctrine and strategy choices is substantially narrowed. The Nunn-Bartlett report observed that "Recent developments in Soviet nuclear force posture, particularly when coupled with Soviet achievement of parity at the strategic nuclear level, suggest that the Soviets are striving to neutralize NATO's tactical nuclear options under "flexible response." During the past several years the Soviets have expanded their nuclear forces in Europe to the point where they may now credibly deter a NATO first use of tactical nuclear weapons...."

The ensuing need to build a capability to mount a more sustained conventional defense, it would appear apparent, should require a substantial change in procurement practices and structure of national forces. But the attainment of priority defense tasks is more likely to run head-on into nationalistic pride, prejudice, and protectionism, bound by domestic political considerations, which, if not overcome, can deny the allied forces the wherewithal they need. Advancing technology and sophisticated equipment are pricing the smaller countries out of maintaining a traditional force structure. All nations tend to focus on the more glamorous and more modern equipment at the expense of more needed mundane expenditures such as, for example, sufficient ammunition or combat ground forces. A realistic and sober appraisal of how best to meet the defense needs is incumbent on all members of the alliance if the most urgent tasks are to be accomplished.

Based on the analysis by our senior military members (Chapter 5), a short-term NATO military priority that needs to be addressed at once is the improvement of the unglamorous conventional fundamentals such as training, supplies, and readiness that have been rationed in the military diet for over a decade. The U.S. military needs to present faithfully its cause for pay raises, modernization, and operational needs. After years of back-burner treatment, the SALT debate and Soviet aggressiveness have caused a national consensus for the need to strengthen our military forces. Now is the time to redress the shortages in unglamorous essentials as well as procurement of new weapons.

Theater Nuclear and Force Imbalance

The second defense policy area deserving concerted political examination concerns short-range military priorities. Four require immediate emphasis, beginning with the issue of long-range tactical nuclear force modernization, an issue that goes to the heart of alliance cohesiveness and confidence. If the political will to modernize these forces falters through Soviet threat or loss of support in the political structure of the alliance, the fragmentation will be severe. The Federal Republic of Germany, under the leadership of Helmut Schmidt, is at the center of this decision. Even in Germany, the decision to accept deployment of the modernized Pershing and the ground-launched cruise missile (GLCM) is predicated on a dual track policy of concurrently seeking negotiations with the USSR to limit theater nuclear weapons.

One short-range politicomilitary priority thus must be to redress the long-range TNF imbalance and seek equitable and verifiable limitation through negotiation. The sensitivity of this issue will require even deeper and more meaningful consultation with our allies than the process used in SALT I and SALT II. It should be clear that even with the decision of December 12, 1979 for 572 launchers, the asymmetrical relationship between Soviet and U.S. capability after 1983 will be adverse. Each of the 572 U.S. LRTNF weapons in Western Europe will have one warhead, whereas the Soviet SS-20 has three warheads, and some 220 have already been deployed. It is fair to state that a European view could well evolve seeking to mitigate NATO's disadvantage without aiming for parity. Under this thesis the risk to the Europeans of "decoupling" would be minimized. Therefore the United States must resolve early to redress the evolving imbalance in theater nuclear weapons. The production of enhanced radiation weapons by the United States would provide a capability for later deployment should a different consensus emerge for strengthening the deterrent.

Early Warning and Moving to Alert Status

As a short-range priority, every effort should be made to enhance our intelligence warning system and our ability quickly to reach coordinated political decisions against impending attack. This is important because the greatest conventional deficiency in alliance ground forces is the lack of forward-deployed active forces in areas of greatest threat. Even those forward-deployed postures in peacetime are frequently 100 or more kilometers from battle stations. If the alliance can have greater reliance on acquiring positive and timely indicators of Warsaw Pact changes in capability for attack, more definitive planning for activation, equipping, and deploying allied reserve forces can ensue. Technology is advancing rapidly in the intelligence monitoring field, but budget constraints are holding back the effort to put this technology at NATO's service. In the United States, Congress and the administration must be pressed to reinvigorate the collection of human intelligence. Concurrently, a vigorous effort to find common ground in confidence-building measures between the NATO alliance and Warsaw Pact forces has some promise and may indeed be the focus toward which nonnuclear arms control efforts in Europe should turn.

No matter how early the warning, in the absence of the ability to respond, word of an impending attack will be of little avail. Although the bureaucracies of the member nations have to some extent identified the constitutional and legal procedures that are required for mobilization and deployment, the political leaders themselves—an ever-changing group of individuals—should become personally aware of the specific nature of the problems they will face should they receive early warning of an impending attack. Personal participation in NATO exercises has proved most valuable to those who have shared that experience.

Such critical times will call for political leadership of the first water in guiding the responses of cabinets, of parliaments, and of popular opinion. Judgments must be made wisely and soundly; they must equally be timely. Governments should be aware that promptly moving to adopt an alert posture in such circumstances, although of course involving important risks, may indeed be the best way to halt external pressures and threats.

Readiness

Finally, as a major short-term military priority, the alliance must undertake the series of short-term military readiness measures outlined in Chapter 5. These include (1) retention of skilled personnel, (2) rapid modernization of equipment, especially defense capabilities, (3) adequacy of sealift and airlift, (4) adequate funds for training, and (5) continuing to redress maldeployment posture of U.S. forces. These measures must be carefully weighed so that priorities can be established. Not all can be funded and supported at the same time. Some are more critical in one NATO country than in another.

LONGER RANGE PRIORITIES

The third overall NATO policy area that merits emphasis concerns longer range priorities. The concept required to establish long-term changes in force structure and capability is heavily constrained by the realities described in Chapter 8 on resources. Nonetheless, efforts to establish a consensus on the increased importance of a graduated massing of conventional resources seems obligatory.

This concept of graduated massing of conventional resources would appear to hold little promise for implementation if suddenly or dramatically thrown at our allies. Even with general recognition of all that this concept entails, it will be a long and expensive process to build the staying power that is necessarily inherent in a shift in this direction. The lines of communication, the ammunition and POL shortages, the spare parts, the reserves, the air- and sealift for U.S. forces, cannot all be dealt with at the same time. This fact gives little cause for early optimism. Political acceptance of the long-term nature of the task is needed from the outset.

Even more urgent than the gradual shift to greater conventional force reliance is the primary objective of establishing a revised infrastructure in NATO, particularly the Benelux countries, Germany, and Italy (and France, in concept) that will create and utilize reserves at an increased level in order to permit the deployment of U.S. forces originally designated for NATO to another region if required. A two- to three-division corps rapid deployment force would have to be compensated for by a similar "ready force" from the reserves of European countries. This should be the shorter term goal. A part of this priority concept is an infrastructure that will enhance the capabilities of national forces to stage through friendly NATO countries. The southern flank of the alliance, particularly Greece and Turkey, takes on an increased importance in this aspect of security priorities. Spain's admittance into the alliance could therefore be important in this regard. Subsequent enlargement of this reserve concept could in the longer term provide for the reinforcement capability ready in place in Europe without the same personnel expenses connected with the standing forces.

In summary, the cohesion of the alliance is a fundamental NATO imperative. Both short- and long-term improvements can be sought under an acceptance of a common threat. None will be easy; most will be politically and economically sensitive. It is true that some may be difficult for political leaders to accept in the 1980s. But real progress must be made in the direction of the improvements outlined above if NATO is to endure as an effective deterrent at a time when deterrence is of the utmost importance.

GLOSSARY

AFCENT	Air Forces Center
AMRAAM	Advanced Medium Range Air to Air Missile
APOE	Aerial Port of Embarkation
ATAF	Allied Tactical Air Force
ATP	Anti Tank Projectile
AWACS	Airborne Warning and Control Systems
CEMA	Council for Mutual Economic Assistance
CENTAG	Central Army Group
CONUS	Continental United States
CRAF	Civil Reserve Air Forces
CSCE	Conference on Security and Cooperation in Europe
CW	Chemical Warfare
EW	Early Warning
FBS	Forward-Based Systems
FEBA	Forward Edge of Battle Area
FEMA	Federal Emergency Management Agency
GCI	Ground Control Intercept
GLCM	Ground-Launched Cruise Missile
GSFG	Group of Soviet Forces—Germany
IRBM	Intermediate Range Ballistic Missile

LOC	Line of Communication
LRA	Long-Range Aviation
LRTNF	Long-Range Theater Nuclear Force
LTDP	Long-Term Defense Plan
MAC	Military Airlift Command
MAU	Military Amphibious Unit
MBFR	Mutual and Balanced Force Reduction
MI	Military Intelligence
MIRV	Multiple Independently Targetable Reentry Vehicles
MOU	Memorandum of Understanding
MPS	Multiple Position Shelters
NBMS	Nuclear Ballistic Missile Submarine
NORTHAG	Northern Army Group
PGM	Precision Guided Missile
POL	Petroleum Oil and Lubrication
POMCUS	Prepositioning of Material Conformed to Unit Sets
RDF	Rapid Deployment Forces
SACEUR	Supreme Allied Commander—Europe
SACLANT	Supreme Allied Commander—Atlantic
SALT	Strategic Arms Limitation Talks
SAM	Surface-to-Air Missile
SLBM	Submarine-Landed Ballistic Missile
SLOC	Sea Lines of Communication
TOW	Tactical Weapon, Optically Sighted, Wire-Guided
USAFE	United States Air Forces—Europe

INDEX

A-7 aircraft (US), 222
A-10 aircraft (US), 123, 146
AAFCE (allied air forces for central Europe), 142
AAMRAM radar missile, 146
Achilles, Theodore C., 209n
Aden, 60, 171, 176
AFCENT (allied forces center), 142
Afghanistan, Soviet invasion of, 13, 14, 57, 60, 62-64, 68, 73, 78, 86, 94, 102, 127, 166, 169, 174, 175, 245
Afrika Corps (Germany), 160
AIM-9L air-to-air missile, 242
Air battle, 120-124
Air-to-air missiles, 145, 242
Air-to-ground missiles, 145
Aircraft carriers, 181, 185
Airlift capability, 155, 173, 181, 183, 185, 187n
Allied Command Europe Mobile Force, 26
ALOC (air line of communication), 149
Angola, 58, 164, 170, 188n
Antiballistic missile (ABM) treaty, 84-85, 252
Antisatellite weapons, 239, 240
APOE interdiction, 123

Arab-Israeli relations, 86, 87, 89, 90, 95, 99-100, 105, 176
Arab-Israeli war (1973), 80, 128-130
Armed attack, defined, 209n
Arms control, 21, 40, 46, 66, 69, 80, 83-85, 94, 132, 167, 213-214. *See also* MBFR; SALT
vs. force modernization, 250-251, 252-253
Arms sales, 100, 174
Army National Training Center, 139
ATAF (allied tactical air force), 142
Atomic Energy Act, 194
Australia, 17, 38, 98
Austria, 116
AWACS (airborne warning and control system), 121, 146, 242

B-1 Bomber (US), 110
B-29, B-50 bombers (US), 193-194
Backfire bomber (USSR), 61, 111, 120, 121, 131, 132, 212
Badger aircraft (USSR), 120, 121
Balance of power, 8, 19-20, 27-28, 39, 59, 62, 63-64, 65, 67, 83-85, 91-92, 107-134, 219-220, 245, 256-257
BAOR (British army on the Rhine), 114

263

Baruch, Bernard, 236
Battle scenarios, 118-120
Belgium 61, 114, 185, 242
Benelux countries, 35, 61, 251, 259
Berlin settlement (1971), 53-54
Bevin, Ernest, 249-250
BMP/BPR/BRDM armored fighting vehicles (USSR), 111
Bohlen, Charles, quoted, 52
Brandt, Willy, 53, 65, 167
Brazil, 101
Bretton Woods, 80, 81
Brezhnev, Leonid, 20, 54, 57, 60, 61, 65, 70
Brown, Harold, 178, 187n
Brzezinski, Zbigniew, 53, 62, 90
Burt, Richard, 166-167

C^3 (command, control, and communications), 125-126
$C^3 I$ (command, control, communications, and intelligence), 219, 239-240
C-5, C-141 aircraft (US), 148, 155
C-5A aircraft (US), 221
Cam Ranh Bay, 171, 176
Cambodia, 170
Canada, 65, 104, 114, 142, 153
Capital investment: and productivity, 234-235
Carter administration, 60, 78, 80-81, 86, 88, 89, 98, 125, 166-167, 170, 175, 212, 251
CENTAG (central army group), 114-116, 141-142
Central America, 13, 16
Chad, 170, 188n
Chemical warfare, 122-123, 141
China, 2, 15, 53, 62, 67, 68, 122
Cleveland, Harlan, 76, 207
Close, Robert, 119
COB (colocated operating bases), 148
Collins, John M., quoted, 180
Colonialism, demise of, 2, 165, 171
Commando units, 185
Conference for Security and Cooperation in Europe, 41, 55, 65, 77, 84
Confidence-building measures, 252, 257
Conscription, 36-37, 47, 137, 139, 238-239

CONUS (continental United States) forces, 137-141
Copperhead laser antitank projectile (US), 128, 220
"Correlation of forces," 9, 192
CRAF (civilian reserve air fleet), 147
Cruise missile (US), 60-61, 83, 85, 110
Currencies, 81-82
CX air transport (US), 183
Czechoslovakia, 54, 64-65, 71, 94, 124, 245
 military forces, 112, 114, 119, 120

"Decoupling," 257
Defuse funding, 35-36, 37, 40, 43, 127, 141, 184-185, 213-215, 245-246
Defense Industrial Reserve Act (1973), 222
Defense industries, 221-230
Defense procurement: rationalization, standardization, and interoperability, 38-39, 46, 47, 126, 213, 221
Defense Production Act (1950), 222
Defense Science Board, 221
DeGaulle, Charles, 197
Denmark, 152
Détente, 9, 18, 19, 39, 51, 54, 55-58, 69, 77-78, 82-83, 87, 89, 167
 and "linkage," 55, 93-94, 252
Deterrence, 198
Deutsche mark, 82
Diego Garcia, 159, 176, 185
Diplomacy, 99-100
Dollar, 81-82
Dragon antitank missile (US), 128
Dulles, John Foster, 194, 195

East-West trade, 18-19, 55-56, 57, 66, 102-104
Eastern Europe, 2, 9, 19, 20, 51, 52-53, 54, 71, 119
Economic aid, 36, 47-48, 95, 100, 174
Economic sanctions, 94, 101-104, 166
Egypt, 189n. *See also* Arab-Israeli relations
82nd airborne division (US), 155, 181
Eisenhower, Dwight, 193, 212

Electronic warfare, 126, 127–128
Enhanced radiation warheads, 110, 257
Environmental Protection Agency, 231
Ethiopia, 14, 60, 78, 86, 164, 170
Euro-Communism, 57, 69
European (Economic) Community, 81, 82, 100, 105
European Monetary System, 81, 82
Exchange rates, 81–82

F-4 aircraft (US), 59, 121, 130
F-14 aircraft (US), 222
F-15 aircraft (US), 121, 146, 155
F-16 aircraft (US), 121, 145, 242
F-111 aircraft (US), 121, 122, 130, 146, 155
FEBA (forward edge of the battle area), 116
Federal Emergency Management Agency, 222–223
Fencers (USSR), 111, 120
Finlandization, 66–67
Fitters (USSR), 111, 120
"Flexible response" strategy, 108, 196, 197–198, 199, 208, 255
Ford administration, 110, 212
"Forward defense" strategy, 23, 74, 75, 85, 117, 133, 193, 196, 198, 203, 255, 257
France, 35, 36, 63, 66, 100, 116, 132, 166, 198
  disarmament proposal (CDE), 21, 252
  and Middle East, 17, 97, 98–99, 104
  military forces, 114, 132, 141, 153, 174, 175
  and NATO, 132, 133, 141, 196, 197, 259
  nuclear forces, 69, 74, 132, 185, 186, 188n

GCI (ground-controlled intercept) sites, 120, 121
German Democratic Republic (East Germany), 19, 53, 56, 77
  military forces, 112, 114, 119
Germany, Federal Republic of (West Germany), 36, 61, 83, 101, 166, 214, 251, 255, 259
  and East Germany, 19, 77
  economic aid, 36, 95, 100
  economy, 35, 82, 213
  and Middle East, 99, 104
  military forces, 61, 69, 74, 75, 97, 98, 114, 116, 118, 142, 153, 185, 242
  US troops in, 64–66, 141–142
  and USSR, 17, 19, 51, 53–54, 56, 63, 65, 66, 83, 84, 103, 167, 256
Giscard d'Estaing, Valery, 82
GLCM (ground-launched cruise missiles), 131, 132, 256
GNP (gross national product), 34, 35, 36, 37, 107
Goldsborough, James O., quoted, 58
Gomulka, Wladyslaw, 54
Goodpaster (SACEUR), 199
Greece, 9, 25, 36, 100, 105, 152, 153, 259
Gromyke, Andrei, 60
Gruenther (SACEUR), 195
GSFG (Soviet forces in East Germany), 119, 124
Guam doctrine (US), 177–178

Haig, Alexander M., Jr., 199, 208
Harmel report (1967), 83, 251
Hawk missile, 243
Hayward, Thomas B., 179
Hellfire laser PGM (US), 220
Helsinki accords, 2, 41, 54, 58, 69, 77
HNS (host nation support) agreements, 240
Honest John missile (US), 108
Human rights, 10–11, 55, 69, 166
Hungary, 52, 54

ICBM (intercontinental ballistic missile), 110
India, 101
Indian Ocean, 172
Indochina, 164
Inflation, 81, 82, 139, 144, 212
Intelligence, 206, 257
International Energy Agency, 101
International Fuel Cycle Evaluation, 100
Iran, 177
  US hostages in, 62–63, 73, 86, 102, 141, 173, 174

and USSR, 78, 86–87, 164, 171
war with Iraq, 64, 87, 88, 99, 153, 169, 170
Iraq, 64, 87, 88, 99, 152, 153, 169, 170, 176, 177
IRBM (intermediate-range ballistic missile), 111, 130, 195
IRR (individual ready reserve), 138, 236–237
Israel, 79, 86, 87, 89, 90, 95, 99–100, 105, 176
　Arab–Israeli war (1973), 80, 128–130
Italy, 36, 104, 105, 153, 185, 259

Japan, 2, 15, 37–38, 62, 67, 68, 97, 104, 105, 154
Johnson administration, 109
Joint Logistics Review Board, 231
Jordan, 64

Kelley, Paul X., 178–179
Kennedy administration, 109, 196, 212
Kenya, 177
Khomeini, Ayatollah, 171
Khrushchev, Nikita, 53
Kiesinger, Kurt, 53
Kissinger, Henry, 84
Komer, Robert, 176
Korean War, 179, 193–194, 225
Kurile Islands, 68
Kuwait, 170, 176

LANTIRN, 145
Lemnitzer (SACEUR), 195, 199
Levy, Walter, 101
Libya, 64, 152, 170, 188n
Lift capability, 155, 173, 181, 183, 185, 187n
Limited-yield accurate missiles, 239
"Linkage": and détente, 55, 93–94, 251–252
LOCs (land lines of communication), 116
LRTNF (long-range theater nuclear forces), 28, 46, 83, 125, 131–132, 167, 202, 250, 252, 256–257
LTDP (long-term defense program), 125–128
　vs. arms control, 250–251, 252–253

M-60 tank (US), 222
M109 howitzer, 243
M113 armored personnel carrier, 221, 243
MAC (military airlift command), 147, 148, 155
MAG-58 tank machine gun (Belgian), 242
Malta, 159
Mansfield amendment, 65
Mao Tse Tung, 62
Marines, 181, 185
"Massive retaliation" strategy, 108, 194, 195, 196, 197
MAU (marine amphibious unit), 154, 155, 173, 181–183, 187n
Maverick air-to-ground missile (US), 145
MBFR (mutual balanced force reduction), 41, 58–61, 65–66, 84, 252
Megatonnage, 131–132
Merchant marine, 26–27, 160–161
Meyer, Edward C., 179
Micunovic, Velko, quoted, 53
Middle East, 7, 33, 62–64, 85, 164
　and USSR, 3, 14–15, 57, 66–67, 78–79, 80, 86, 88–89, 92, 93, 152–153, 169–171
　and Western allies, 16, 17, 32–34, 57, 66–67, 79–80, 86–105 passim, 152–156, 166, 175–183 passim
Military balance, 8, 19–20, 27–28, 39, 59–67 passim, 83–85, 91–92, 107–134, 219–220, 245, 256–257
Military production, 221–230
Minuteman missile (US), 110
Mobilization time, 118, 119, 257–258
Modernization programs, 131–132, 146
Monetary and trade relations, 80, 81–82
Mozambique, 164, 170
MPSs (maritime pre-positioning ships), 183
Multiple launch rocket system, 243
MX missile (US), 110

National Defense Act (1920), 225
National Guard, 138, 237
NATO, 3–5, 8, 9
　Article 6, 209n

conventional defense forces, 22–31 *passim*, 40–47 *passim*, 75, 110, 112–120, 194–195, 202–203, 219–220, 243–244, 255–256
crisis management, 205–206
Defense Planning Committee, 197
defense procurement, 38–39, 46, 47, 126, 213, 221, 241–243
"the Fourteen," 197
logistics, 157–162
and MBFR, 58–61, 65–66
Middle East forces, 98–99, 152–156
military committee documents, 194, 195, 196, 197, 198
northern and southern flanks, 24–26, 46, 152–153, 259
nuclear "strategic umbrella," 27–31, 75, 194, 201
readiness, 136–147, 258
reinforcements, reserves, and resupply, 26–27, 32–34, 44, 45–46, 91–92, 98–99, 117–118, 123–124, 128, 147–151, 214, 259
sea power, 172, 181–183
security threat outside Treaty area, 12–17, 85–87, 96, 97, 104, 163–186 *passim*, 253
strategy, 116–117, 133–134, 191–208 *passim*. See also "Flexible response" strategy; "Forward defense" strategy; "Massive retaliation" strategy
and Third World, 172, 173, 174–175
Natural resources, critical, 34–35, 164, 165, 203, 222, 225
Netherlands, 61, 114, 118
New Zealand, 38
"Nifty Nugget" exercise, 228, 233
Nixon administration, 178, 212
Norstad, Lauris, 195, 199
North Asia, 2
North Korea, 15
North Vietnam, 62
NORTHAG (northern army group), 114–116
Northeast Asia, 15
Norway, 25, 35, 79, 152, 153
Nuclear energy, 79, 100–101
Nuclear nonproliferation, 79, 100–101
Nuclear Nonproliferation Act (1978), 100

Nuclear storage and stockpiles, 122, 195
Nuclear strategy, 27–31, 75, 194, 201
Nuclear test ban, 84
Nunn–Bartlett report, 119, 255

Occupational Safety and Health Administration, 231
Office of War Mobilization, 225
Oil, 153–154, 164, 166, 203
Oil embargo (1973–1974), 164
Okinawa, 159
Olympic Games boycott (Moscow, 1980), 166, 173–174
Oman, 99, 170, 177, 185–186
101st air assault division (US), 155
OPEC (Organization of Petroleum Exporting Countries), 79, 101, 102, 164
Organization for Economic Cooperation and Development, 235
Ostpolitik, 84, 93, 103

Pacifism, 20–21, 83
Pakistan, 36, 68, 169, 177
Palestine Liberation Organization, 57
Palestinian issue, 87, 89, 99. See also Arab–Israeli relations
Perry, William, 179
Pershing missiles (US), 59, 60, 83, 85, 130, 131, 132, 256
PGM (precision-guided missiles), 128–130, 220
Poland, 36, 119
  Soviet threat to, 19, 41, 52, 54–55, 71–72, 78, 94, 103
Polaris submarine (UK), 130
POMCUS (prepositioned materiel configured to unit sets), 118, 145, 147, 150
Poseidon warheads, 130
PWRM (prepositioned war reserve materiel), 149–150

Qatar, 170

Radio Free Europe and Voice of America, 22, 48
Ranger battalions (US), 155
RDF (rapid deployment force), 91, 98, 154–155, 175–176, 179–183, 214, 259

Reagan administration, 105, 137, 167, 184, 212, 222, 251, 252
Realpolitik, 93, 94
Recession, 57, 80
Research and development programs, 221
Reserve currency, 81–82
Reston, James, 62
Rogers, Bernard W., 111, 198, 199, 200, 206
Roland missile, 243
Romania, 54
Rommel, Erwin, 160
RSI (rationalization, standardization, and interoperability), 38–39, 46, 47, 126, 213, 221, 241–243

SA-6, SA-8 (USSR), 121
Sabah, Sabah al Ahmad al, 176–177
Sabotage and terrorism, 118, 122
SAC (strategic air command), 193–194
SACEUR (Supreme Allied Commander—Europe), 195, 198, 199
SACLANT, 154
SALT (Strategic Arms Limitation Treaty), 59, 60–61, 78, 84, 252
SAM (Soviet surface-to-air missile), 128, 146
Saudi Arabia, 64, 100, 170
Schlesinger, James R., 163, 174
Schmidt, Helmut, 57–58, 59, 61, 63, 82, 131, 256
Schmuckle, Gerd, 207
Sealift capability, 155, 172–173, 181, 183, 185, 187n
Shaba, 170, 185
SHAPE (Supreme Headquarters Allied Powers Europe), 198
Sidewinder air-to-air missile (US), 145
Sino–Vietnamese war (1979), 68
6th air cavalry combat brigade (US), 155
Slay, Alton D., 228; quoted, 232, 247n
SLOC interdiction, 123
"Snake," the, 81
Socotra, 171, 176
Solidarity (Poland), 71
Somalia, 177

South Asia, 62–64
South Korea, 2, 9
Southeast Asia, 13, 16
Soviet Union, 2, 9, 51–72 *passim*
 and China, 53, 62
 conventional forces, 111, 112–120
 defense funding, 240, 247n
 and Eastern Europe, 20, 51, 52–53, 54, 71
 economy, 56, 71, 102–103, 240
 encirclement, fear of, 60, 62, 67, 86
 leadership succession in, 68, 70–71, 245
 and MBFR, 58–61
 and Middle East, 14–15, 78–79, 80, 88–89, 92, 152–153, 168, 169–171, 176
 military buildup, 22, 77, 119, 120–121, 158, 164, 181, 196, 212
 and NATO's flanks, 25, 152–153
 nuclear forces, 29–30, 130–132, 212, 255
 and Third World, 12–13, 14, 15, 16, 42–43, 60, 62–64, 78, 86, 92, 164, 167, 168–171, 200
 and West Germany, 17, 51, 52, 53–54, 56, 58, 64, 65, 77, 103
 and Western Europe, 17–19, 51, 53–55, 64–70, 89, 167
Spain, 198
Sparrow air-to-air missile (US), 145
SS-4, SS-5 IRBMs (USSR), 130
SS-20 mobile IRBM (USSR), 28, 59, 61, 111, 131, 132, 212, 257
Standing Naval Force Atlantic, 26
Stockpiling, 145–146, 223
Strait of Hormuz, 166
Strategic nuclear weapons, 109–110
STYX naval missile (USSR), 129–130
Submarine-launched ballistic missile, 132
Suez Canal, 158, 165
"Surge" capacity, 233–234
Switzerland, 116
Syria, 64

T-72 tank (USSR), 111
Tactical air support, 114–116, 121–122, 123
Taiwan, 2

Tank guns, 242
Tanks (USSR), 112–114
Technology transfer, 19, 102–103
Thailand, 13, 16, 170
Theater nuclear weapons, 108–109, 111, 130–132, 255
Third World, 2, 101, 165
  and USSR, 8, 12–13, 39–40, 78, 81, 92, 164, 168–171
  and the Western allies, 13–14, 16, 36, 42, 75–76, 85, 164, 172, 173–175
Tomahawk missile, 131
Tornado aircraft, 121
TOW antitank missile (US), 128, 221
Trident submarine (US), 110
Tunisia, 188n
"Tunnel," the, 81
Turkey, 9, 25, 36, 46, 95, 100, 152, 153, 259
24th mechanized division (US), 155

U-2 (US), 129
United Arab Emirates, 170
United Kingdom, 17, 61, 79, 97, 98–99, 100, 104, 165, 213, 251
  defense white paper (1980), 171
  military forces, 114, 130, 185
  nuclear forces, 69, 74, 153
  and Oman, 99, 170, 185–186
United States, 3, 54, 206
  Air Force, 137–138, 145, 188n
  Army, 137, 142–144, 181, 188n, 215, 236–239
  Bureau of Labor Statistics, 224, 234–235
  CONUS forces, 128, 137–141
  Defense Department, 141, 222, 228
  defense funding, 35–36, 37, 40, 43, 141, 184, 189n, 214, 240, 244, 247n
  economic sanctions, 101–104
  economy, 80, 81, 212, 214
  global containment policy, 75, 78, 79–80, 97, 214
  hostages in Iran, 62–63, 73, 86, 102, 141, 173, 174
  industrial mobilization, 221–230
  and the Middle East, 17, 40, 98–99, 104, 154–156, 175–183 *passim*
  military bases overseas, 176, 177, 189n
  military forces in Europe, 64–66, 114, 141–147, 214
  military reputation, 173–174, 178
  Navy, 180, 187n, 188n
  nuclear energy, 100–101
  rapid deployment force, 91, 98, 154–155, 175–176, 179–183, 214, 259
  recommendations for, 46–48
  reserve units, 117, 138–139, 236–237
  and the Third World, 75–76, 173–174
  and West Germany, 214
USAFE (United States Air Force units in Europe), 145

Vandenberg, Arthur, 4, 5
Vietnam, 16, 68, 170
Vietnam War, 3, 8, 65, 81, 88, 129, 160, 174, 178, 180, 225, 231, 234
Vulcan bomber (UK), 130

War Production Board, 225
Warning time, 117–118, 151, 240, 257–258
Wars, "short" and "long," 228–229
Warsaw Pact–NATO comparisons, 180, 181, 256–257
  equipment, 146–147
  megatonnage, 131–132
  mobilization, 118, 119
  personnel, 238
  resources, 34–35
  theater nuclear weapons, 130–132
Weather factor, 121–122, 145
Weinberger, Casper, 154
Western Europe, 2, 86, 251
  defense funding, 35, 36, 37, 40, 43, 184–185, 213, 214, 244, 245–246
  economy, 80, 92
  and the Middle East, 16, 97, 98–99, 154, 166
  and the Third World, 13–14, 16, 36, 164
  and United States, 10, 18, 40, 55–57, 58, 63, 69, 73–106 *passim*, 94–98, 99–104, 166–167

and USSR, 17–19, 51, 53–55, 64–70, 89, 167
"Window of vulnerability," 110
Wohlstetter, Albert, 166

XM-tank, 221, 242

Yemen, 62, 78, 86, 164, 170

Zaire, 60, 188n
ZSU-23/57 radar-guided antiaircraft gun (USSR), 111, 121

UA
646.3
C73
1982

OCT 1 1982

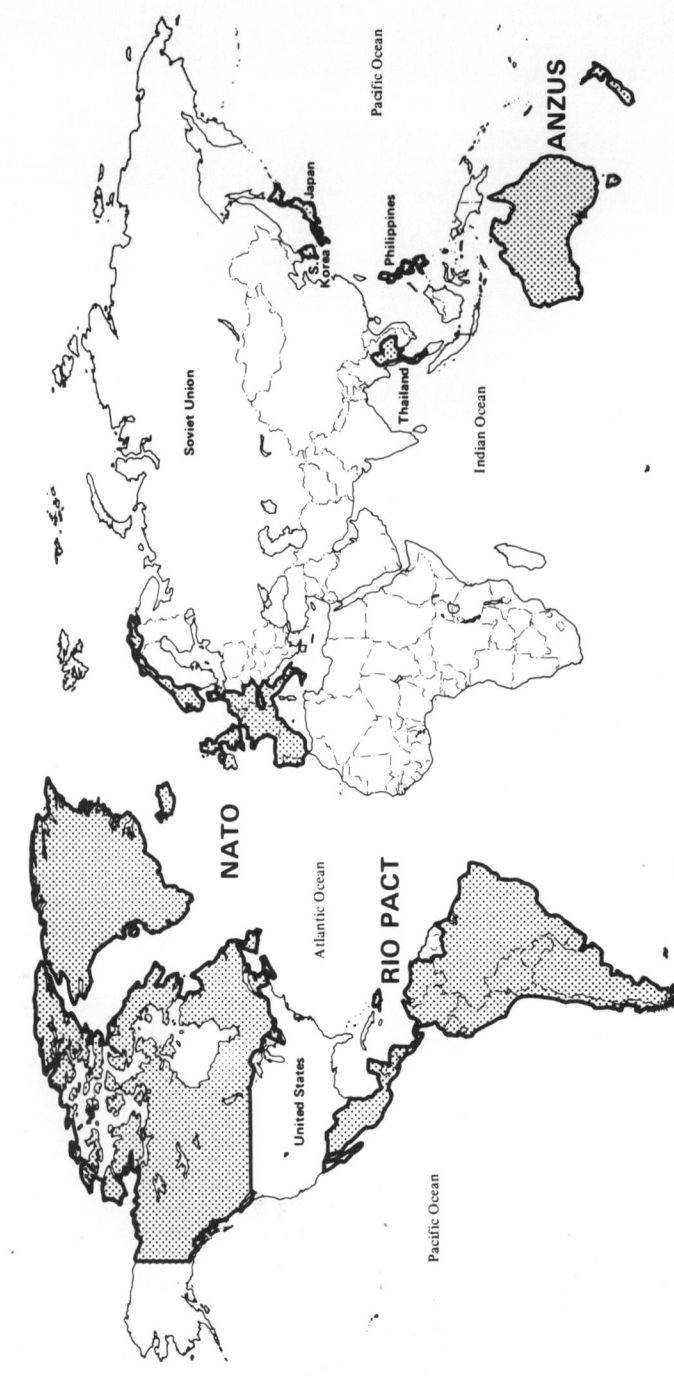

U.S. Security Treaties

Note: Permission granted by the Library of Congress to reproduce this map from the publication entitled "U.S.–Soviet Military Balance Concepts and Capabilities 1960–1980."